THE BEAT OF LIFE

For Josef and Olivia

REINHARD FRIEDL

For Tex Poole (1926–2020),
whose noble heart gave out aged ninety-three

GERT REIFARTH

DR REINHARD FRIEDL

WITH SHIRLEY MICHAELA SEUL
TRANSLATED BY GERT REIFARTH

THE BEAT OF LIFE

A surgeon reveals the secrets of the heart

Published by Hero,
an imprint of Legend Times Group
51 Gower Street
London WC1E 6HJ
hero@hero-press.com
www.hero-press.com

Originally published as *Der Takt des Lebens. Warum das Herz unser wichtigstes Sinnesorgan ist* by Reinhard Friedl, with Shirley Michaela Seul
First published in English in Australia by Black Inc.,
an imprint of Schwartz Books, in 2021
This edition first published in 2021

This book is intended for informational purposes only. It is not a substitute for professional medical advice, diagnosis or treatment. The experiences of patients described in this book have been fictionalised.

9781800316058 (hardback)
9781800316065 (ebook)

Cover design: Ditte Løkkegaard
Text design and typesetting by Typography Studio
Printing managed by Jellyfish Solutions Ltd

'You always hear of people who lost their mind because of love. But there are also many who lost love because of their mind.'

Jean Paul, *novelist*

'I had explored the mysteries of the brain, and it was time to devote as much academic rigour and hard science to exploring the secrets of the heart.'

James R. Doty, *clinical professor of neurosurgery at Stanford, and the founder and director of the Center for Compassion and Altruism Research and Education*

CONTENTS

CONTENTS

REVEALING THE HEART

Ba-Boom, Ba-Boom, Ba-Boom

Most of the time you don't hear it, but if your heartbeat were suddenly to stop, you'd stop too. You live from one beat to the next. In between, death resides. If after one heartbeat there isn't another, the clock of life stands still. It might happen while we're sleeping – or shopping. None of us knows the hour of our death.

Your heartbeat is my profession. Sixty to eighty times per minute, this sound creates life. Most hearts beat calmly and strongly, some in a constant rush. Even if the heart stumbles occasionally, it always tries to go on. I have seen many hearts labouring with their last ounce of strength. The heart knows no weekend, no holiday. On your seventy-fifth birthday, it will have beaten about three billion times. It started its work eight months before your birth, twenty-two days after procreation. The heart is the first organ to develop, long before the brain and the first breath. Nothing works without the heart. It throbs through the years and decades, unnoticed – until something ceases to function. Or until a high-tech scan, by accident, discovers a defect that has not yet been felt.

Afflictions of the heart are always dramatic. A pain in the chest is completely different from a pain in the hip. We perceive everything to do with the heart as an attack on our lives, on our inviolability. Even if later it turns out not to be life-threatening, an aching heart is a cause for concern and often triggers a fear of dying. A headache can also be a harbinger of danger; it can eventually lead to death by stroke or brain haemorrhage. Yet a severe headache worries us less than a light pressure in the chest. Deep inside, human beings sense that the heart is the source of all life.

As a heart surgeon I have held thousands of hearts in my hands. I have operated on premature babies and repaired the heart valves of patients well advanced in years. I have implanted artificial hearts and stitched up knife wounds to the heart. As an organ, the heart has been investigated down to its smallest parts. We seem to know everything about it – and yet we know nothing. Every week there are hundreds of new scientific findings published about this organ that has not changed since *Homo sapiens* emerged 300,000 years ago.[1] It seems that the French philosopher and mathematician Blaise Pascal is still correct: 'The heart has its reasons of which reason knows nothing.'

Independently, separated by time and space, without knowledge of each other and despite different languages, all human beings draw hearts to express love, both earthly and celestial. Does this point to an inner truth anchored deeply in every human being? Or merely to a desire that we all share unconsciously? All great cultures, from the Stone Age to the present, and all religions and spiritual movements perceived and continue to perceive the heart as a symbol, as the biological centre for love, compassion, joy, courage, strength, truth and wisdom. But in the age of heart transplants and data migration, the magic of the heart seems to have vanished – as if it could not withstand our mechanised world. But maybe those qualities are precisely what we need for a more humane future.

In Saint-Exupéry's *The Little Prince*, the fox says: 'One sees clearly only with the heart.' Yet so far we have not found eyes on our biological heart, nor sensors for compassion and love, nor a pump discharging courage and strength. However, we all experience these heart qualities as an inner reality that guides our lives. How, then, are these related to our physical, pumping heart? What can science reveal about this 'other' heart and its dimensions of consciousness? And how do these things influence illnesses and therapy?

Aristotle believed the heart rather than the brain was the source of all emotions. Modern neuroscience argues that love originates in the brain. Has it stolen the secrets of love away from the heart? And is our language only a memory – but of what? Or is it just trivial metaphors when we speak of someone becoming close to our heart, of closing off our heart to someone or inviting them into it, of losing or taking heart, of a weight felt on or being taken off our heart, of a broken heart or something that makes our heart stop, of stealing someone's heart or giving ours away, of wearing our heart on our sleeve, which is better than something saddening it or making it sink? What is dear to the heart's heart? Some of these symptoms are indeed taken to a cardiologist – for example when they manifest as cardiac arrhythmia or angina pectoris. *Doctor, I feel as if there's a stone weighing on my chest.* In the past I took care of these people solely as a surgeon; today I am interested in the whole human being.

Heart surgeons can put hearts to sleep and make them beat – but they usually don't speak with a lot of heart, rather in terms of mechanics: heart-lung machines, ECG, ultrasound or even artificial hearts. And of course they speak with their colleagues – assistant doctors, anaesthetists, cardio technicians, surgical nurses. A heart operation is not an intimate matter. The heart, hidden deeply in the chest and well protected by the ribs, is opened up to the glaring light of high-tech operating theatres under the concentrated gaze of many pairs of eyes. To the heart surgeon, it is first of all a pump they have to repair, the motor of life. In contrast to all other doctors, heart surgeons don't only know this motor's functions through video images and data generated with the help of ultrasound or computed tomography (CT) scans, intracardiac catheters or magnetic resonance imaging (MRI). Even in this era of high-tech medicine, it makes a big difference to one's understanding of this organ to see it with one's own eyes and touch it with one's own hands – rather than just observe it second-hand via monitors.

As a heart surgeon, I reach deep into the chest and put hand to heart. A heart is not used to such contact. Hearts can react very sensitively to touch. Some get a fright and respond with arrhythmia. Yet even sick hearts are strong, so powerful that their inherent strength astonishes me time and time again. When they lie in my hand it feels as if they are the essence of life, the pure and absolute will to live. For me, every heart is its own being; every heart has its own appearance. I never know what is about to be revealed to me when I cut the skin with a scalpel and open the chest. Some hearts are very lively and muscular, others a little chubby with clearly visible fat. Some betray their long path through life and appear tired and spent. Yet they all have one thing in common: they like nothing more than to beat.

What is it, really, that I am holding in my hand? Merely a pump, or the origin of all human consciousness?

Neuroscience has not yet been able to explain the mystery of how consciousness arises. The prevalent thinking suggests it emerges as a result of biochemical and electrophysiological processes in the central nervous system and the brain. Neuroscientists know the components, their functions and complicated circuitry very well. However, how something intellectual emerges out of the organic matter of our bodies, how a thought or an emotion is formed – this remains largely unknown. According to famous neurosurgeon Eben Alexander and other brain researchers, the emergence of consciousness is a blank spot on the map of neuroscience.[2] What if the heart could fill in at least part of this unknown territory?

Shut down

I remember when I saw a heart for the first time as a young doctor. It reminded me of a tender, freshly peeled fruit. I looked at the orange-sized organ with awe. Pumping muscles, partly covered by a thin layer of fat, nothing more. Or so it seemed to me at first. I had to hold the suction drain to take away leaking blood and was glad to have something in my hand to hold onto – the sensations were so overwhelming. The heart continued to beat unflinchingly while my colleagues prepared for the heart-lung machine to be connected to it and placed numerous sutures on the heart and the aorta.

Most heart operations can only be carried out after shutting down the heart; this is done by interrupting the supply of blood and therefore oxygen. To achieve a protected shutdown of the heart, a specific mix of liquids, mainly blood and potassium, is infused into the arteries. In this way, the heart's electrical activity is shut down – it stops beating. Its energy consumption is thus reduced and its cells need less oxygen. Sometimes it is also cooled. The heart can then survive for a certain time without major damage, until it is supplied with blood again. In order for this artificially created cardiac arrest not to lead to death, the heart is connected to a heart-lung machine before being shut down. This machine now transports the blood (instead of the heart, which takes a break during the operation) and also supplies the heart with oxygen.

Often someone needs to hold the heart in a certain position during an operation, for the surgeon to have access to the heart chambers while operating on the valves and to be able to fix a bypass to the posterior wall. This task usually falls to the youngest assistant doctor. So suddenly a heart was placed in my hand. I did not know what was happening and stood stock-still. *Let me not damage anything, let me not make a mistake.* I held the heart as I would a premature baby. It seemed so delicate, so fragile, even though simultaneously I could feel its strong muscles, even in its shut-down state. I did not yet have any sense of its consistency and texture, its essence. I think it was a man's heart, but it could have been a woman's – they look nearly the same, are the same size and strength and both weigh about 300 grams.

As the months and years passed, my tasks as a prospective heart surgeon in the operating theatre became more demanding. I was allowed to open the chest, to take over the connection to the heart-lung machine, eventually to suture my first bypass. My teachers supervised my every move with watchful eyes, occasionally even led my hand. I was so nervous that I sometimes nearly dropped the needle clamp, and my own heart nearly stopped whenever the smallest drop of blood leaked through. *Was I too deep or still okay?* My experienced colleague did not say anything, so I continued. I learned to control my emotions, or better still not have any. Millimetre by millimetre the needle crept through the parchment-thin walls of the arteries whose inner diameter is only a millimetre or two. *If the aorta, life's great stream, were to tear, the blood could squirt up to the operating light . . .* this thought makes the hands of every beginner shake. A clear mind and utterly precise mechanics make an operation safe. I internalised this truth and became more confident, while outside the operating theatre, this need to function without emotion left a mark on my heart – and in my unconditional drive to become a good surgeon, I did not even notice that.

Eight years later, now a heart surgeon myself, I sutured the skin after a long operation without thinking much about it. I was in awe of technology and its fascinating possibilities. I had learned how to conduct a certain kind of bypass operation routinely, even on a beating heart. But during most operations the heart is shut down, as described above, then repaired and started again. Both the complex repairs of valves and the enormous responsibility of my profession excited me. The heart does not easily forgive mistakes, and the time pressure is enormous, as the shutdown of the heart should not normally last more than sixty to ninety minutes. The shorter the better. It is a bit like a tyre change in Formula One; the pit stop can't be too long or the patient will be out of the race. I had a wonderful teacher who put it like this: 'The patient must not skid off the road during an operation, and this requires a motor that has been repaired in a technically excellent fashion.'

As a doctor, naturally my first desire was to help my patients. But I won't deny that the outward glory, the admiration that is often bestowed on heart surgeons, also appealed to me. Their renown and the responsibility they shoulder are similar to that of a jet pilot. But compared to the heart, a plane is a rather straightforward machine, one that always reacts in a similar way due to predictable technical laws. However, as a heart surgeon I cannot rely on any definite causality such as 'When I press this button, the valve will shake'. The mechanics of the heart are much more subtle and uncontrollable – it is not unusual for hearts to behave differently from how one expects them to. Surgeons have to anticipate anything at all, have to stay in control, remain calm and must never allow themselves to be led by emotion. All of which I perfected.

Restarted

Day in, day out, I worked with hearts. My life existed in the sterile reality of the operating theatre. My encounters with hearts were restricted to standing in front of opened-up chests every day. With my team I brought half-dead patients back to life and repaired pumps to restore their owners' quality of life. I rarely thought beyond the operating table. Heart operations take place at the well of life, which afterwards starts to gush again for most people. But some die. You have to put that aside. You must not feel too much, compassion least of all, otherwise you will cease to function. One day I realised that I was no longer able to hear the voice of my own heart properly. More and more I asked questions that had nothing to do with surgery. *Is the heart more than a pump? Can we sense things consciously with the heart? Maybe even act from the heart? Is there a connection between the voice of the heart and illnesses, between a fulfilling life and one full of suffering?* I wanted answers to all these questions. You hold in your hands the result of my journey into the secrets of the heart.

Where to begin? As a scientist I researched the work of colleagues from my own field and then went beyond my discipline. Mathematicians, engineers and heart specialists increasingly approach the heart with technology and virtual reality – which is extraordinarily fascinating. I developed cardiovascular navigation systems myself. But unfortunately I did not find what I was looking for. One day a bright red heart beamed at me in a train station's bookshop. 'Heart Ache' read the headline of the *Bild* newspaper, followed by the question: 'How do I know if my heart is sick?' That was something I wouldn't mind knowing myself, so I read more closely and found the article contained only well-known facts about heart attacks and related matters. A magazine next to it promised 'Everything you need to know about high blood pressure, cholesterol, heart attacks, vascular constriction, heart failure, arrhythmia, angina and organ replacement'. As a scientist, you shouldn't sit in an ivory tower, so I invested eight euros and ninety cents – and learned nothing new.

A friend sent me an article: 'Neurology: How the gut rules the head'. In it, I read interesting things about the gut's nervous system and its communication with the 'head brain', as it was called – and I learned that anger affects the stomach and that certain types of consciousness originate in the intestines. 'The brain in the gut is highly intelligent,' declares American neuroscientist Michael D. Gershon, chair of the Department of Pathology and Cell Biology at Columbia University. If there was 'intelligence' in the gut, hitherto known only for digestion and producing faeces, there could well also be something to discover higher up, in the heart. But reading on, I received quite a blow: 'The heart, in contrast, is a primitive pump,' explained Gershon.[3] For a moment, my heart was knocked down. But it refused to be out for the count; it beat in my throat, up to my brain. I felt outraged. Primitive? My heart? Never!

I used to be a boxer and had learned to anticipate my opponent's punches. This one, though, caught me flat-footed. Somewhere deep inside I knew it wasn't true. But how could I prove that? I held the heart's biology, its sounds and mechanics in high esteem – like a violin maker who loves his Stradivarius: a violin is not merely a wooden box with four strings. A universe of music and emotion emerges when someone knows how to play it.

If the heart was only a primitive pump, how had nobody managed to manufacture a replica that worked even nearly as perfectly? Why could one not simply replace the heart? Why did so many people die waiting for a transplant?

My warrior heart awoke. In a sense I went back to square one and left behind the safe ground I had acquired as a heart surgeon. I started to ask questions that reached beyond the operating table. I looked anew at this 'primitive pump', from unconventional perspectives, in order to reconcile it with my experiences in thousands of heart operations. In the operating theatre my eyes became more watchful. I did not want to miss the slightest movement of the heart.

Over and over during this time I noticed discrepancies between medical and figurative accounts of the heart. I talked to others about this. 'The heart is only ever allowed to be good,' an acquaintance once objected after I had listed a few expressions about the heart. 'But there is also a cowardly heart, a cold heart, an ice-cold heart, one made of stone and a frozen one.'

At first, I was speechless. She was undoubtedly right; instead of dismissing her point, I had to address this issue. Had I fallen victim to my own longing? Then I thought of a sick heart, one no longer properly supplied with blood, which had thus become cold. Its fear, having extracted all courage from it, had let calcium turn it to stone. I knew suddenly that her objection confirmed my sentiment. The heart itself is strong, life-affirming – positive. Fear is stagnant flow, frozen energy.

Sometimes it is necessary to take a step back in order to see the whole picture. As a scientist I learned that successful women and men, great personalities, always use heart *and* mind. It is not only permitted, but indeed our essential purpose, to take nothing for granted, to question everything, to think anew. It is from the realm of the inconceivable that science draws its most interesting questions.

In the following pages, I reveal the miracle that is the heart. I explain how I arrived at my findings and how they may change our lives for the better. The heart is with us from our mother's womb until we rejoin Mother Earth at the end of life. We speak about it so often, but this essential organ is nevertheless strangely foreign to many of us, and some are even afraid of it. Some people are glad when they don't feel their heart, as that must mean everything's okay, mustn't it? That is what I, too, thought for a long time – until I was proven wrong. Today I believe medicine should acknowledge that the heart is not merely a pump but a source of life and consciousness and also plays a role in many disturbances and illnesses; this conviction would grant us deeper understanding of the body's logistics and lead to healthier and more joyful lives.

While searching for the heart's true nature I also rediscovered my own heart. It seems that health professionals lean towards a certain heartlessness, perhaps believing that we need to close our hearts off because we are confronted with so much suffering. Society rewards us for this with recognition and high status, preferring the ice-cold surgeon to the compassionate one whose hands may shake. Today I know that my compassionate hands do not shake: no, they have eyes. The holistic reawakening of my heart has altered my life dramatically, changing the way I approach myself and relate to my fellow human beings and of course my patients. They are not vessels whose pump is on strike but holistic human beings on whom I bestow sympathy as well as medicine.

Nobody would deny that love exists. But can we measure it? We can sense and feel inner phenomena that we can't measure objectively. We can't verify their existence in a lab or completely explain their molecular formation. But we sense them intuitively. More and more people look for answers, maybe because they 'sense' that a life based solely on technology and acceleration does not make us happier or more content. Is it possible to return to our selves and into our hearts? Body-oriented therapies and spiritual movements try to give people awareness of their body and thereby 'open their hearts'. Something seems to be missing from our comfortable, modern but also performance-oriented and increasingly mechanised lives. Our Western way of life, so rational and yet hardly sensible, does not correlate with the idea of a holistic existence.

In the following chapters I tell the story of a heart surgeon who set out to rediscover his missing heart. I have compiled both age-old and up-to-date findings from various scientific disciplines, especially concerning the secret connections of heart and brain and their influence on emotions and consciousness. This will result in new insights for the health of the whole heart and its therapy – for the whole human being.

Listening to the heart

Ba-boom, ba-boom, ba-boom – the healthy heart makes two sounds, of which the first one sounds a little shorter. It is not a monotonous military march, not a *boom-boom, boom-boom*; rather, its shortened first sound creates a light, dance-like rhythm. We can hear the *ba-boom* if someone allows us to put our ear on their chest. As a child I loved to listen to the heartbeats of my siblings and parents. The first sound, the 'ba', is created when the heart contracts and the valves open; the second, the *boom*, comes when the heart relaxes and the valves close. We know from embryology and prenatal diagnosis that this is the first sound we hear in our lives, long before we are born. It is immensely important for the relationship between mother and child.

Being allowed to listen to another person's heart is a very intimate and special experience. Listening to a heart requires closeness, even in medicine. 'Please take off your shirt.' What happens after that – I hear your heart! – is a diagnostic measure that is used less and less often. In earlier centuries and decades it was considered a doctor's greatest skill to be able to distinguish the fine tones and sounds of the heart and make the correct diagnosis. Even when I was training, one of the most highly regarded medical journals, *The New England Journal of Medicine*, published an article detailing how auscultation of the heart – experienced experts listening to it – can result in highly accurate diagnoses.[4]

Thirty years later it is sobering for heart surgeons when a patient tells them immediately before an operation: 'You are the first doctor to listen to my heart.'[5] For most patients, a doctor listening to their heart represents the highest intimacy between medical professional and patient. I strongly believe this also points to a deep desire of human beings for their inner sanctum to be listened to.

Today the moving image has replaced sound. I have observed that cardiologists these days listen to their patients' hearts only rarely. I mean not only the tones and sounds generated by the heart's mechanics, but also the heart's other voice, which comes from the spheres of wisdom and compassion. Heart researchers, psychologists and spiritual teachers increasingly agree: this voice is inextricably connected with the organic heart.

When did you last hear it?

I want to invite you to consciously perceive your heart. Put your right hand on your heart and feel your heartbeat. Breathe steadily and perceive what your heart has to tell you.

Whatever it is you hear or don't hear, it is right. Don't lose your patience and don't judge. There are many different kinds of meditation. The aim is always the same: pause, breathe more slowly, sense what's inside and observe the present and all thoughts and feelings.

You may feel, after a while, how your heart gives you life with every *ba-boom*. Now. From moment to moment, endowing life with new vitality.

When I practised this meditation for the first time more than ten years ago, I heard nothing. I was at the very beginning of my journey to the true heart. Back then, the heart was still a pump to me, albeit not a primitive one but one that was a miracle of precision and strength.

THE SIX-CYLINDER BIO-TURBINE

Every day the heart transports 9000 litres of blood and beats about 100,000 times while doing so. The heart's large muscle strands twist around each other like spirals and form fantastic cave systems connected by valves and separated by septums. Inner and outer layers of muscle rotate simultaneously in order to discharge blood, following a finely synchronised choreography.[1] Physically, you could therefore view the heart not merely as a pump but almost as a turbine, an engine. And if you sometimes feel as if your heart is jumping in your body – that, too, is understandable. If you see it when the chest is opened, it jumps rather than runs. A heart reveals a lot about its condition to a heart surgeon through these jumps and turns and their rhythm. The geometric shapes of this dance of the heart are as unique as a fingerprint. You could use them for identification, as there are no two people in the world with an identical heart. Researchers from the University at Buffalo used this discovery and developed the prototype of a heart radar. In the future people may log onto their computer using their heartprint, use it to pay in the supermarket or check in at an airport and prove their identity with their heart.[2]

Usually, though, the heart likes to hide from overly curious examination. For heart surgeons to be able to see it, they first have to gain access to its engine room, which is the pericardium (also known as the heart sac). The pericardium forms a wet cave for the heart to glide in without becoming sore. It supports the heart without constricting it. It also prevents it from becoming too full – and helps it remain where it belongs in stressful or energising situations. There have been instances, after heart operations, where the heart has herniated into the lung cavities – a deadly complication if it is not immediately recognised and corrected.

If the heart has been injured or has survived inflammation, its home may form a hard shell. First, pericardium and heart stick together, which is known as pericarditis, or inflammation of the heart sac. Later, calcium deposits massively restrict the movements of the heart; it is no longer able to fill properly and becomes hopelessly trapped in a stranglehold. Its once-cosy home becomes a deadly trap, the custom-made suit a straitjacket. This situation is fittingly called an 'armoured heart' (constrictive pericarditis).

Armour has no place around the heart – it impedes rather than protects it, trapping it in many ways. Biologically, this means the heart loses its ability to receive enough blood. This can lead to death.

A healthy pericardium, then, protects the heart and provides it with its own space in which it can beat freely, without impediment or constriction. The heart needs both protection and freedom. In traditional Chinese medicine, the heart sac is called 'the mother of blood and protector of the heart'. Our academic medicine validates this view, as the pericardium also improves immunity and protects the heart from infection. The versatility of the highly flexible and simultaneously tear-proof pericardium is a huge help for heart surgeons during an operation. Occasionally we even use a tiny piece of it during our operations; a few square centimetres of 'pericardial patch' (as heart surgeons call it) are extracted, for example, to repair a valve or close a hole when treating an innate heart defect.

The heart functions as a pump, and in order to transport blood, its base and tip move towards each other in a smooth motion. Its liquid freight is conducted through labyrinthine caves – polymorphous and fantastically formed cathedrals of life featuring muscular pillars, distinct ledges and fine sinews. They provide stability and strength and generate the difficult mechanics of the six heart valves, which ensure that the blood in the cave system flows forwards and never backwards.

Yes, you read that correctly. There are six valves, the smallest of which have the longest names: *valvula thebesii* and *valvula venae cavae inferioris* (which feature mainly in high-level questions in anatomy exams). They help pass on the low-oxygen blood that is sucked into the right atrium. The bloodstream must not stop. Patients become very ill if valves are leaky or severely calcified, or if there are faulty connections.

In a healthy heart, the body's low-oxygen venous blood flows from the right atrium through the tricuspid valve into the right ventricle; from there it is passed on to the lungs, which have a capacity of several litres. Breathing now comes into play. Every time we breathe out, the lungs draw carbon dioxide from the blood and release it into the air. When we inhale, our red blood cells are charged with oxygen atoms. Thus refreshed, the high-oxygen blood is now sucked into the left atrium and is ready for the big trip into our body's most remote cells. With the next heartbeat, it will pass through the mitral valve, flood the left ventricle in a flash and then leave the heart through the aortic valve.

The left ventricle is a muscular powerhouse which builds the pressure we feel in the arteries as our pulse. Blood circulates in a closed circuit. This was first discovered only 360 years ago by English doctor and scientist William Harvey. Previously, for more than 1400 years, doctors had followed the erroneous teaching of the ancient Roman physician Galen, who claimed blood is generated in the liver and trickles from there to other organs. He viewed the heart as a kind of oven in which a flame burned to cleanse the blood; the smoke was said to be discharged via the lungs. Harvey had the courage to doubt the predominant doctrine of his time – and was ridiculed by many. However, he laid the foundations for our modern understanding of body and heart; it is thanks to him that today we know blood circulates and returns to the heart via the veins.[3]

Thus, the heart not only gives – it can simultaneously take. With every contraction, blood is not only pumped out but also sucked in. Giving and taking is the heart's most fundamental principle, and also the essence of love and balanced interpersonal relationships and partnerships. This muscle receives and distributes eight to ten tons of blood every day in an adult. Imagine you had to carry nine tons of blood across the street and back again. That would be 900 ten-litre buckets. How long do you think you would keep going? The heart goes one better still: it doesn't just 'carry' your blood across the street but rather pumps it through a blood vessel system that is 100,000 kilometres long. All of this happens with every heartbeat inside your body, and yet we rarely think of the heart's downright supernatural accomplishment.

Perhaps you wish to pause again at this point and place a hand on your chest's left side? Don't be shy! You can sense what human beings have felt since the dawn of their evolution. Something alive dwells within you: your heart. It creates life of its own accord. Simultaneously it is at home in your whole body, linked to every single cell. The heart is the mother and the well, whose bloodstream supplies all cells with vital ingredients. And it is the launching pad for the cells of our immune system, without which we would quickly die from infections.

A lively heart is deeply rooted in the body, and its blood vessels are connected to every single cell – the many thousands of billions of cells in our gut, genitals, arms, legs and sensory organs. These vessels reach deep inside us, all the way up to our brain, which they surround with millions of fine and ever finer branches. Unfortunately, this is not recognisable in the completely mutilated depictions of hearts we see in anatomical drawings and health magazines. They remind me of pictures of trees whose roots and crown have been cut off. If you want to get a picture of the whole heart, you must not separate it from its pulsing roots and crown. An unobstructed, strong, powerful stream of life and blood flows inside it. It is animated by electric energy; its highly energetic electromagnetic field is measurable over a metre away from a human being.[4] Sometimes it seems to me that we humans understand far more about trees than about our own hearts.

With every contraction of the heart a pulse wave rushes through our arteries, and the brain pulses continuously to the beat of the heart. No organ is as dependent on the heart-driven blood flow as the brain. If this flow is interrupted for more than a few seconds, we faint.

We know the pulse mostly from the media and TV shows such as *ER*, *House* or numerous crime thrillers. 'Is the patient still alive?' And then the obligatory fingers to the throat to check for a pulse, and sometimes the call: 'Quick! Get a doctor!'

HEART ON THE TABLE

It is a bone-chillingly cold winter night. I turn into the clinic's car park twenty-nine minutes after receiving the call. The large windows of the operating wing are brightly lit, and I know my team is making all the preparations necessary for a life-saving intervention. The connection where the aorta emerges from a patient's heart threatens to tear. It is a race against time, which had already begun when the phone call woke me in the middle of the night.

'Are you awake?' the assistant doctor asked me.

'Yes,' I replied, already alarmed and on my way to the bathroom.

There are two types of calls. The more pleasant one starts with: 'Good morning, nothing serious, I just have a question.' 'Are you awake?' calls do not usually bode well. And so it was with this one: 'We are about to receive a Type A dissection; the helicopter will land in fifteen minutes.

The patient is sixty years old, I don't know any more yet.'

'I'll be there in thirty-five minutes,' I said. If such an emergency is not operated on, mortality increases by 2 per cent per hour in the first twenty-four hours. In other words, 2 per cent of patients die within the first hour, 4 per cent within two hours and 48 per cent within twenty-four hours.

In the operating theatre I familiarise myself with the case. I learn from the emergency doctor's report that the patient had mentioned an extreme pain in the chest that afternoon. This excruciating pain is likened to an elephant sitting on someone's chest, and it often reaches into the back. The farmer had been working in the forest felling trees. It took a while until the emergency doctor reached him, and he was admitted to the emergency department of a clinic thirty kilometres away with a suspected heart attack. More precious time passed until it was established he was not having a heart attack. The images from the CT scans are unmistakable; the aorta is enlarged to nearly double the normal diameter, now measuring eight centimetres. It is separated by the clearly visible dissection membrane, a rag of tissue that now hangs on the inside of the aorta like part of a burst tyre. With this illness, the innermost layer of the body's biggest blood vessel will tear and the powerful bloodstream from the heart will use the wrong channel.

The patient has to be operated on immediately, as now only a paper-thin wall stands between life and death. It won't withstand the enormous pressure for much longer. Blood is already seeping through the remaining dam, and soon it will break. In the pericardium (the heart sac) there is already a large amount of blood, pressing on the heart. The patient's heart needs to be relieved if we are to save his life, as he has a condition known as cardiac tamponade. As the wrong channel is being used, the heart is choked by its own blood, which fills the pericardium. Even a strong heart cannot bear this for long. And this one is not very strong. We watch the drama on the ultrasound monitor that provides a live visual of what is happening. Beads of sweat appear on the assistant's forehead. The nurse looks at me questioningly. The experienced anaesthetist informs me that the aortic valve is leaky and may have to be replaced or repaired.

'But that doesn't matter at the moment,' I reply gruffly, even though she is right. It really does not matter now. My reaction betrays my own nervousness. The patient's blood pressure is low and he is beginning to go into shock. Life or death. I need to concentrate on the most important issue.

'Knife,' I say curtly to the nurse and cut the skin above the sternum. A thin stream of blood follows the scalpel's straight incision. The pressure on the blade is just strong enough to cut the skin but not yet the tissue underneath. There will be a long scar above the heart, from the throat to the end of the sternum; it will mark the patient for the rest of his life. How long that will be I don't want to predict right now. The only thing that is clear is that the patient – of whom I only see a small section, as he is covered in green cloth – is critically ill. With a further cut I advance to the sternum. Wordlessly, the nurse passes me the pneumatic surgical saw. Its howling sound drowns out the diminishing bleeps of the heart rate monitor. Someone hastily fetches the sternal retractor – a kind of jack that will hold open and prise apart the bony ribcage. Deep down I see the heart sac, inflated close to bursting, full of blood gleaming a dark shade of red that suggests danger.

'Scissors,' I say. When everything is on the line, you make do without 'please' and 'thank you'. I open the heart sac. If the aortic wall doesn't hold now, it is highly likely the patient will bleed to death. But if I don't open the heart sac, he will have a cardiac arrest. We have to act immediately.

I feel nothing. With the calmness that always comes over me when I operate, I do what needs to be done. My movements slow down for a moment and I look into the eyes of my assistant. He is experienced. I like and trust him. He has often been by my side during difficult operations and has helped me enormously. He knows what is at stake and what will happen now. He already has the big suction drain in his hand. I nod, he nods back. All of this happens in less than a second. Then I cut into the heart sac, and immediately half a litre of blood squirts out. We can no longer see anything. It is like looking into a bath of blood: you can't make out the bottom beneath all the muddy, bloody liquid. You don't know how dangerous it is in there or what lurks below. The suction runs on maximum, and finally I see the heart again. This organ, usually so majestic, is crouched in an unnatural, deformed position in the left corner of its home, the heart sac. Normally the heart is proudly enthroned in the middle, but now it has been pushed aside by the enlarged aorta.

'Pressure rising,' I hear the anaesthetist say. I take a deep breath; the strange icy-cold feeling in my body subsides. The aorta has held and the lethal threat has been averted. For now.

I prepare the connection to the heart-lung machine. The whole team knows that we are about to undertake one of the biggest, most complex and most dangerous operations that can be carried out in heart surgery, or on a person in general. The extent of the operation is not yet clear. That, too, is particular to this illness; even the most advanced imaging processes cannot tell you which parts of the aorta and arteries, potentially leading to the brain, are affected. The only thing that is clear is that it will take many hours – and the outcome is unknown.

As the aorta is damaged, the heart-lung machine will be connected via the subclavian artery below the collarbone. This artery is now exposed; parallel to it a wire is pushed into the heart via a large vein in the groin – under the stern gaze of the anaesthetist, who follows the procedure on ultrasound and tells me if my wire is going the right way.

'Thirty thousand units of heparin,' I command. This is to make it impossible for the patient's blood to clot.

'Thirty thousand units of heparin administered,' I hear shortly afterwards. I like these clear and concise calls. It is like in a space station or an airplane. Deep concentration, without frills – the nitty-gritty.

The cannulas are inserted, the heart-lung machine is started. 'Full flow, six litres,' reports the cardio technician. That reassures me immensely. So far we have been acrobats on a high wire. Every complication could have been deadly. Now we have installed a parallel circuit, enabling us to return to the patient the blood we suck away and to shut down the heart at any time. But we aren't quite there yet. First, I fully open the heart sac and assess the damage. The aorta, life's magnificent and powerful channel, lies bloated and bloodshot in the opened chest. We don't know much about the patient, but his blood pressure has been too high for years and the connecting tube from heart to body continued to expand unnoticed and without the patient being able to sense any symptoms – until the innermost layer tore. The question is only where exactly the tear is and how much tissue has been damaged. Does the tear reach all the way to the brain arteries or down to the coronary arteries? Sometimes a tear can also come close to a heart valve, namely the aortic valve, so it also has to be replaced and repaired. At exactly the point where it emerges from the heart, the aorta is a little wider. Leonardo da Vinci knew that this widening, called the sinus of Valsalva, swirls the blood in a symmetric and even harmonious fashion. Even today the purpose of this swirling is not fully understood, but it improves the profile and properties of the flow. Possibly there is information encoded in these swirls which the heart sends to the body and the brain.

However, one thing is clear. This pipe needs to be replaced, and that can only happen with a complete shutdown of the cardiovascular system. So if the patient holds on, there will be a point in the next few hours when not only the heart pauses but the heart-lung machine as well. That is the unique feature of this particular operation. My instruction is: 'Cool down to eighteen degrees of rectal temperature.' So using the heat exchanger of the heart-lung machine, the blood is cooled down progressively until the body's core temperature in the bowel is 18°C. Why there? The temperature taken in the bottom is the true temperature, as it is measured within the body, and is more reliable than measuring the skin temperature under the arm. It takes at least an hour to cool down a person by lowering the temperature of their blood from 36°C to 18°C. I use this time to lay open the aorta and its connections to the brain and wrap them in sterile cloth strips in order to mark every single blood vessel and thereby gain an overview of the anatomy. Although in principle most people follow the same blueprint, we are vastly diverse creatures. Just as two identical faces are extremely rare, the way we are arranged internally is also individual. I tread with the utmost care, as if the area around the aorta were a minefield. One careless move and the life-threatening situation, which we have somewhat under control at the moment, will escalate.

At about 26°C, ventricular fibrillation sets in. This would be the cause of death of someone suffering from hypothermia after falling into a crevasse or drifting in the Arctic Sea. In our case this reaction is intentional. While my gaze is fixed on the aorta and my vision rather restricted by glasses with a magnification of two and a half, I receive my information from the cardio technician, who has his eyes constantly on the monitors: 'Heart fibrillating.' I look past the magnifiers and see it too. Even though this situation has been deliberately brought about, a fibrillating heart is still a picture of misery. Ever since witnessing my first heart operation some decades ago, I am always affected when a heart fibrillates. It looks so miserable, evoking my pity: the last twitches of a dying creature. Of course I know the heart is not a creature, but in some way it seems like one. It reminds me of a bird that has flown into a windowpane and is now twitching to death. The incomparable beauty of a heart's movement and its natural power are laid low. With this operation, however, the shutting down of the heart is part of the unavoidable process and part of its success. I request a twenty-five-centimetre-long aortic clamp and put it on the aorta with the utmost accurateness and care. I close it millimetre by millimetre until a soft click indicates it is locked in place. Every sausage has two ends. The clamp is sitting in the middle of the sausage.

The end towards the heart, including the heart itself, is no longer supplied with blood.

From now on we fight even harder against time. Even at these low temperatures, the heart's blood supply should not be cut off for more than ninety minutes, if at all possible, in order for the heart muscle not to become damaged. I open the aorta close to the heart with a scalpel and see inside it for the first time. The surface of the inner wall, formerly baby-skin smooth, is now fissured like a rocky potato field. Life's most important vein is corrugated by many centimetres of calcium. The tear is clearly visible: the innermost layer, called the endothelium, is hanging down like torn-off wallpaper in a house that needs restoration. Temporary stitches are put in, as it is important to remain visually in control and be able to clearly see every corner. The tear ends a centimetre and a half above the coronary arteries, and the aortic valve has also not been affected. One could say the patient is lucky. But will his heart continue to serve him? We inject a mixture of blood and potassium into the coronary vessels with special cannulas. The heart stops twitching and fibrillating. The desired heart shutdown has been achieved. In this state the heart will need the least amount of energy and oxygen. Neither are available at the moment. The heart will have to hold its breath – and hang on.

At one end of the sausage we will need to attach a new pipe. I measure the size for a prosthetic tube. The tissue to which it needs to be attached is fragile like wet toilet paper, so I have to sew in felt rings to strengthen it. I am familiar with this kind of tailoring from my mother. She used to sew strengthening material into my torn pants so the patches would hold. Leaky seams will lead to nasty complications. But only many hours into the operation will I know if everything is really tight. I stitch with the utmost diligence and concentration. Even the tiniest tear, a slightly-too-big distance from one stitch to the next, can become a rapid leak. The seam is a success and we approach the epicentre of the operation. The patient's body temperature is at 18°C. His head is resting on ice to protect the brain.

'Head further down, suction ready, stop machine,' I say, this time really commandingly. The patient no longer has any circulation, the heart-lung machine's rotation pumps also stand still. His head is lowered so no air will penetrate the head vessels. You could call his state clinical death. Everything will have to work, not a minute or even a second must be wasted. The total cardiovascular arrest should last forty-five minutes at most so the brain is not damaged. This technique is only possible because the brain's demand for oxygen is very low at 18°C – which is why accident victims sometimes survive in ice-cold water. Because of the intense hypothermia, they are not brain dead and the heart can resume its work.

I open the aortic clamp and look into the part of the aorta that leads to the brain – the other end of the sausage. You can't look any deeper into a human being. It is very still in the theatre. Usually during an operation, the patient still has circulation when I look into their resting heart. Not this one. His connection to life has been shut down – hopefully only temporarily. Medical team and patient are on a path through the no-man's-land between life and death.

Meticulously, we inspect the interior of the aorta and its branches leading to the brain arteries, one of the most important nodes in the human blood vessel system. The heart gives 30 per cent of its blood to the brain, even though the brain accounts for only 2 per cent of body mass. Without this massive subsidy of energy from the heart, our high-performance processor and server room would break down immediately. Under normal circumstances the brain continuously tells the heart how much blood and oxygen it needs, and the heart's pumping is essential in maintaining this cerebral circulation.[1] Even if the heart is not well, is in shock or is massively damaged, it will try to maintain blood supply to the brain, at the expense of the skin and other organs. If there is such a thing as unconditional love, this is pretty close.

The aortic arch looks better than I feared. No further tear is visible. So the connections to the brain will not have to be repaired and implanted separately. I remove as much of the damaged aortic tissue as possible and support the rest with felt strips. I carefully measure a number of times and cut the end of the prosthetic tube accordingly. The tube must not be too short but must not bend either. Felt strips are attached again, and the tube is now sewn on at the other end too, with utmost care. Blood rich in oxygen is injected through special cannulas in order to further protect the brain. Before finishing the seam, the whole system will have to be vented thoroughly to avoid damage to the heart and brain from air pockets.

Simultaneously, the heart-lung machine is started slowly and the aortic clamp released. This is a decisive moment as now it will be revealed whether the seams are tight or there is any bleeding. At this stage in the intervention, the patient's blood is severely affected. It is reminiscent of tap water – there is not the slightest clotting, it drips and leaks everywhere. One has to trust that once clotting begins things will improve. But that might take a while.

We have now been operating for three hours and will continue to do so for another three at least. We start to warm the patient; he will be at his original body temperature of 36°C in about an hour. At the moment there is nothing to prepare. It is four o'clock at night, or in the morning . . . and we wait. Everyone is exhausted, their faces grey, their eyes bloodshot. A leaden fatigue clings to the ceiling and wants to descend on us. I tell my assistant to go and have a coffee, and lethargically watch the temperature indicator, which is climbing torturously slowly. The heart has started to fibrillate again and is being defibrillated at 28°C. For this purpose two metal spoons, which look like salad servers, are placed on the heart, and an electric shock is administered. The heart twitches briefly and continues to fibrillate. We try again. Nothing. Third time lucky – the heart is beating again. And even though we are completely exhausted, we are glad. We're immediately awake again. The patient's beating heart, even though it appears to be a bit disoriented (at least that is my interpretation), impresses us with its vitality. There is nothing more beautiful than a beating heart. After all these years I am still deeply moved by it, time and time again. It is still a damn miracle. I love being able to see it so closely.

Low conversations begin, someone tells a joke. We don't know if the patient will survive the operation and what the damage will be if he does. But his heart is beating, which is a good start. And the beat is a strong one. It appears to rejoice with us.

The red-haired anaesthetist slams her metal stool down in front of the table, climbs up and observes the result of the operation critically. She calls the green cloth that separates the patient from the area of operation the 'blood–brain barrier'. The theatre is amused. Still, we have left behind a bloodbath. Which is what the anaesthetist is mercilessly pointing out with grand understatement: 'It's still a bit wet. I have RCC, FFP and PC at the ready. Tell me when I should use them.' These are blood and clotting agents which have been generously donated by others: red blood cells, plasma and platelets. The patient will need them in copious amounts.

We sew on the pacemaker cables and wean the patient from the heart-lung machine; the heart is working well. Here and there more stitches are needed where the increasing blood pressure reveals leaks. Then we remove the cannulas and stitch up the holes. After more than six hours of operating and administering forty bags of blood products, clotting is still very limited. Such an operation is also a battle of material: it produces several bags of waste.

We did what we could. Now the patient has to recover. We leave him open, which is to say the chest is provisionally stuffed with sterile fabric and closed with self-adhesive film. Sometimes this procedure is repeated over several days – until the patient's own clotting mechanism has been restored. Until then he will remain in intensive care under anaesthetic in a critical condition.

THE BLEEDING HEART

I was around eight years old and I stared at the altar in fascination. I liked to accompany my grandma to the Maria-Hilf chapel near Roggenburg, Bavaria – just a ten-minute walk – to see the heart. It was Mary's. The heart was very red, very fleshy, and she held it in both hands in front of her breast. It looked as if it held a secret, and I found it revolting. But I was fascinated at the same time, so I stared at it, mesmerised, while my grandma knelt down beside me and prayed. On the way home, she would always tell me the same story from the Thirty Years' War, which she could recount as if she had been there. Her 'we' didn't just include her family but our whole village. Soldiers had plundered and pillaged our village. A brave priest, the last survivor, remained in the monastery disguised as a peasant. 'But on the hill where the chapel stands today,' grandma's voice became softer, 'they caught him and hanged him from an oak tree.' Now she whispered with reverence: 'But Mary came to his aid and stopped the rope from strangling him.'

I saw Mary in front of me. She flew down from the sky, her hair billowing, her red fleshy heart in front of her chest. Today I am astonished that such a sexual depiction would grace a place of pilgrimage. Religion, too, holds many secrets.

Death by hanging occupied my thoughts quite a bit when I was a boy. At grandma's I was allowed to watch Westerns, and somehow the rope featured in many of them. In almost every Western, someone was hanged from a tree, often the only one on the prairie. But they were always saved at the last moment. I knew who was behind that, even though she sent her messengers in the shape of Indians.

On butchering day, I looked forward to the heart, which was what I most liked to eat. It came fresh out of the large bubbling pot and was cut up; everyone standing close by snatched a piece.

One day, my mother gave me a book about Our Lady of Marienfried, a place of prayer in Pfaffenhofen. There she was again, the Virgin Mary, this time appearing to a pious, desperate girl. Was everyone in danger of being haunted by Mary? I did not want to experience an apparition of Mary. I did not want to be offered a red, fleshy heart. If any situation became dangerous in this regard, I would squint my eyes firmly – I was mortally afraid of seeing Mary approach me with her bleeding heart. At Claretinerkolleg, where I went to school, the Madonna was also depicted with an exposed heart, pierced by a few arrows. Sometimes she only had a red spot on her chest; I preferred that. The community of Claretian monks call themselves 'Sons of the Immaculate Heart of the Blessed Virgin Mary'. Bare-breasted, bare-hearted, merciful. Claret, the order's founder, was a missionary in Cuba. Five hundred years ago Spanish conquerors had brutally massacred the Aztecs of Central America. They, too, had strange customs: they cut out the hearts of live human victims and held them up, still beating, to their sun god. The Aztecs possessed astonishing surgical abilities, as it is by no means easy to cut out a heart so quickly that it is still beating after the violent act. You definitely can't go through the lengthy procedure of opening the chest. Instead, the Aztecs would cut the skin and muscles at the bottom edge of the left ribcage and grab the heart from below. It has to be done swiftly. Then the heart can indeed beat for a short while without a brain or a body. It is autonomous; I have known this since my childhood.

My grandma lived on a farm. Her hens ranged freely. In the evenings, we kids were allowed to look for eggs. I often found some in the barn behind the diesel barrels for Grandpa's tractor. We also loved to chase the hens, but we never caught one. Grandma did, though. She snatched her victim, put it on the chopping block and severed the head with an axe. Once she held the hen by the head instead of the body and watched, its head in her hand, as the hen ran away. It made it to the vegetable garden. I was thoroughly impressed, and for a long time thought of this hen whenever the word 'headless' was used. You don't get far without a head. Nor without a heart.

Heart hierarchy

The headless chicken died and was eaten on Sunday. That was completely normal to me. I was not afraid of it – in contrast to my fear of apparitions of Mary. The heart's true nature, and why the hen had been able to run without a head – that was what interested me. I don't believe that as a child I ever wanted to help Mary or other Madonnas with a heart operation, or to close their open chests. I studied medicine and in my doctoral thesis wrote about multimedia simulations of heart cases; this was subsequently published as one of the first multimedia CDs. The connection between technology and medicine really fascinated me; the first computers came on the market when I was a student – my Atari and I were bosom buddies. But after some time at the German Heart Centre in Berlin I realised I would not be happy as a cardiologist, even though the work would involve a lot of technology and computer simulations. I wanted to get to the real, true heart. After a stint in Munich as a general physician, I eventually reached my goal: I was appointed as an assistant doctor in the field of heart surgery.

At various points in my career I focused on my key competence, surgery, and improved at it – which I noticed as my operations became more complex. The difficult cases, too, those which are risky from the outset, were now entrusted to me: emergencies, patients advanced in age, those who'd had a stroke or those whose pumping function was generally bad.

For decades I saw the heart exclusively as a pump; most heart surgeons probably do. Sure, there was a patient attached to the pump, but in the operating theatre I saw only the orange square of the sanitised chest, surrounded by sterile cloth. Man or woman? It didn't matter. The main thing was to repair the pump. As one of my colleagues used to say: I am a heart surgeon because I prefer patients under anaesthetic.

I agreed with him because in that way I was best able to concentrate on the work. The 'rest' could be dealt with by psychologists and other doctors.

I know a pathologist who put it even more chillingly: 'I prefer cold patients to warm ones.' For a patient not to end up on the autopsy table, a heart surgeon has to learn to remain cool even in precarious situations.

Double heart surgeon

I found conversations with relatives after very difficult operations unpleasant. What could I say? The patient was alive after all. The next few days would show if he or she would survive. It is possible, with today's highly technical medicine, to wheel nearly every patient out of the theatre alive – if necessary, connected to a heart-lung machine that provides something called ECMO, or extracorporeal membrane oxygenation. Some hearts just need time and recover in the next few days, but many patients die in intensive care. Every patient is assessed to determine their risk group. A surgeon who 'produces' a disproportionate number of fatalities – which will become evident via the national quality assurance statistics – will quickly be withdrawn from the operating theatre. Thus, statistics breathe down your neck – and that's a good thing.

You can also get a pain in the neck when operating, by the way. Most people have no idea how physically demanding it is to be a surgeon. You stand leaning slightly forward at the operating table, sometimes for seven or eight hours at a time. Without eating, drinking or going to the bathroom – you can't simply let a patient lie with an open chest, connected to the heart-lung machine, while going to the canteen to order a beetroot smoothie. You learn to defer your needs, and become conditioned to this after a time. In the theatre, there is a supply of liquids and blood only for the patient.

But it is worth it. Heart operations have helped millions of people. Most of them recover and can enjoy life again. Yet despite these marvellous results, heart surgery is a challenging profession – consider bullying, humiliation, stress, high responsibility, overtime, complaints, teasing, unpleasant bosses, lots of bureaucracy, the need to prove one's worth and, oh yes, patients and relatives. For years I preferred conversations with patients to those with their relatives. Both groups want to know the truth. But relatives still hope for miracles, while patients sometimes instinctively know how things stand with them. In the end, everyone is grateful for the truth.

I never found it hard to calmly and clearly discourage a patient from having an operation. But since I have recovered my lost heart, since I am no longer merely a heart surgeon but a double heart surgeon, so to speak, a lot has changed for me during such conversations, even though their content has remained the same. I do not sugarcoat anything: 'You don't stand a chance. You will remain in intensive care for a long time, and you won't be well. It is more than unlikely that you will ever be able to return home again, and if you did you would need assistance around the clock. If you were my father or my grandmother, I would not advise you to have this operation. I would tell you to spend time with your loved ones and to make any important arrangements you still deem necessary.'

Such conversations are exhausting, and intense. In them, too, I touch my patients' hearts. I do not need to open a chest to do so, nor do I need any instruments. But I myself must open up a little as a human being, unlike during an operation. I was lucky to have a few role models who encouraged me to speak the truth. There are many factors involved. If a patient is high-risk but desperately wants to explore their slim chance, one can proceed with the operation – but it requires a frank talk beforehand. I can remember cases where a so-called high-risk patient who was fierce and motivated survived the first thirty days. This is an important milestone in heart surgery. After that time the first big hurdle has been overcome. However, it often still takes a year or more before a heart operation has been fully 'absorbed', physically and psychologically.

The wise heart

One day a patient gave me a book. I often received little gifts from my patients: flowers, a bottle of wine or books – usually ones to do with gratitude or the heart. This one I remember particularly clearly. It was *The Wise Heart*, by American psychologist and spiritual teacher Jack Kornfield, who is famous for promoting Buddhism in the West. Normally I do not read such books. But I browsed this one a little – and became engrossed. What strange, fascinating, intriguing thoughts! They all revolved around the heart, which in these pages did not bleed or need oxygen or require connection to a heart-lung machine – no, it was wise. As if it were more than a pump. As if it were the source of qualities such as love and compassion – heart qualities, so to speak. These had so far been limited in my mind to data such as pumping function and blood pressure.

The book awakened something in me. Wanting to know more, I registered for a meditation workshop. There were about a dozen of us, more women than men, sitting on the floor in a light-filled room. It smelled of incense and we drank tea. Just like at my workplace, lots was said about the heart – yet in a way I had not experienced before. I had no idea which field these specialists belonged to.

The leader of the group asked us to open our hearts.

I thought: *Without anaesthetic?*

We were asked to breathe into our hearts.

Without a lens tube?

We were asked to feel our hearts.

Feel? The heart? Meaning mine? I made an effort and realised that I was hungry. What else was I meant to feel?

The others in the circle seemed to feel a lot, judging from their faces; they looked as if they had been intravenously given a millilitre of diazepam. I, on the other hand, was sober. And hungry, as I sensed in my upper abdomen. But also curious – because even though we all used the same language, I had no idea what they meant. Yet I was a heart surgeon. It was me who was the specialist here. Thousands of hearts had been in my hands.

At the end everyone was asked to say what they felt.

'I am deeply moved right now,' I heard. And: 'I feel very connected.' Or: 'I am in touch with a great sorrow, in touch with the child I abandoned.'

Well, I was also in touch. With lunatics, I thought. I treated the case as incurable and told the truth when my turn came: 'I'm hungry.'

'That's a good sign,' the leader replied, without any trace of anger but instead with warmth in her eyes.

And in that moment I realised: I was hungry for the truth that lies in the pump. That was what had been bothering me for months. That's what life is often like, after all – changes don't occur out of the blue but announce themselves long before you can put them into words. And suddenly you know. Because you feel it. Because the heart makes itself known.

ONE HEART AFTER ANOTHER

Sometimes it starts with a visit to the dentist. Or with the flu. Germs like to take up residence in the heart valves, and this frequently leads to serious heart inflammation. It may destroy a valve completely, which in turn makes an operation necessary. Heart valve inflammation can also have other causes. A young man had had his nipples pierced and they had become inflamed. Bacteria travelled into his blood-stream and attacked a heart valve. Now he had been admitted to intensive care.

I already had two complicated operations behind me that day. One artificial heart, one bypass. Routine matters, normally, and the bypass patient was going extremely well. But the implant patient was suffering from bad bleeding after the op. That was not unusual; heart operations are among the bloodiest procedures, because in order to perform them you need to stop the blood from clotting (otherwise it would clot in the heart-lung machine). However, most blood is not lost but caught, cleaned and infused back into the patient.

In the late afternoon – I had just sat down for the first time that day to have a bite to eat – the intensive care unit paged me. The artificial heart patient was still bleeding. My colleagues were unsure if it was due to a clotting problem or a surgical issue. If it was the latter, I would have to operate again. Disrupted clotting can be tackled with a unit of stored blood or an anticoagulant. At the same time an assistant doctor called to tell me the relatives of the patient with the implanted heart were waiting for me, distraught. There were five of them, and the patient's very pregnant wife was at the end of her tether. She wanted to know how the operation had gone.

At that moment I could only tell her that the pump was running, that her husband had bled a lot and that we would have to see how things developed. I would have liked to avoid the conversation, but it was part of my job. On my way to meet with the relatives I received another call from intensive care. The patient with the piercing who had heart valve inflammation had rapidly deteriorated. It was feared he would not survive the weekend. I decided to look after him first.

'The nipple piercer is on five.' A nurse showed me the way. What may sound humorous was far from being funny. The patient was in a really bad condition; he was suffering from sepsis and an operation was both unavoidable and very dangerous. We had to act quickly.

Have I mentioned it was a Friday? It is fairly common for critical patients to be transferred to heart surgery on Friday afternoons because colleagues from other departments fear they will not survive the weekend. Sometimes these colleagues will say: 'Otherwise he might fall off the perch.' While I don't know where this expression comes from, I never want to be the one to blame if that happens. As a heart surgeon I have seen enough dead people, and maybe we develop rather strange ways of talking as we come to terms with such things – humour helps. Somehow. Or cynicism. We line up to save lives, and we fight for every patient until the end, even though we do not know them, perhaps have not even spoken with them before the operation. This patient, however, I would now get to know personally – and I have never forgotten him.

The nipple piercer

Despite his weak condition, the piercer looked like a surfer. Long blond hair, very white teeth, a smile that would have delighted all the female nurses and some of the male ones – if he had been healthy. But now all colour had receded from his face; it was grey. His eyes had retreated far into their sockets. In his current condition one could only pity him. A colleague briefly explained to me that there had been no improvement despite intense antibiotic treatment, first by the family doctor and then at our clinic. An ultrasound had revealed inflammation of the mitral valve, on which I noticed bacteria and wart-shaped bits of destroyed valve tissue. They could peel away at any moment, travel to the brain and cause a stroke. In this way the heart can affect the brain. The valve was partly destroyed and leaking, blood was flowing back to the lungs, which were full of water. I could even hear this – the young man was wheezing and gasping for breath, but he did not want to put on the oxygen mask while he was talking to me. He even tried to smile.

The patient had to be operated on immediately. Even though it was Friday. And I was tired. It could well be a severe intervention. The heart tissue was likely to be inflamed, which meant the 'anchorage' for the new valve would have to be built with pericardium patches first. Why did he have to get that piercing?

Luckily he was not a young woman. I could implant a so-called artificial valve made from carbon or titanium. If all went really well, he would keep it his whole lifetime. An artificial heart valve is a foreign body; blood clots on its surface, which leads to strokes if the clots peel away. Therefore the patient would have to take blood-thinning medication for the rest of his life. A woman on blood-thinning drugs should not become pregnant as there is the danger she might bleed to death while giving birth. So for a female you would implant a so-called organic heart valve made from pig or cattle tissue, to be replaced later with an artificial valve. An organic valve is also chosen for older patients of both genders, as these patients are more prone to injury – for example if they have a fall – and they tend to forget their medication, which can lead to big complications. Also, organic valves last longer in elderly patients than in younger ones, as mechanically they are strained more by young and active people. Despite all attempts, medical technology has not so far succeeded in constructing the ideal heart valve – one that works as perfectly as the one made by Mother Nature. Every type of valve replacement has advantages and disadvantages.

I explained to the patient: 'You have an inflamed heart valve. It's life-threatening. We have to operate. It is a dangerous procedure, but you don't have a choice if you want to survive this. Do you agree to the operation.' I used my usual script.

The young man only looked at me. I read incomprehension in his eyes, and fear. I looked at the clock on the wall. The emergency team would start to operate in an hour. But they were already dealing with another case. My kingdom for a cup of coffee. The young man still did not say anything. Maybe he thought he was dreaming; who would ever expect something like this. Fit as a fiddle, and then out of the blue a heart operation. But I seemed to be dreaming too, because I did something that wasn't at all like me. I took the young man's hot, weak hand, which lay limp on the blanket, in mine. I had never done such a thing before. I touch hearts, not hands. Now his eyes changed. He was looking for an answer. I remained silent and absorbed his question. It travelled through my eyes into ... my heart.

And my heart began to talk. 'If I was in your place, I would be afraid too,' I said. 'A difficult operation lies ahead of you. But you are young and strong, and your heart wants to live. Every heart wants to live. Your heart will do its best, and I will do my best, and together we will make it, okay?'

I read the 'Yes' in his eyes before he said it. And although that was all I had wanted to know, I remained seated beside him. Even though I had things to prepare. Even though I still hadn't had any coffee. I remained holding his wet, feverish hand – and if the leader of yesterday's workshop had asked me right then to feel my heart, I would not have had to search in vain. My heart had woken up.

CHIMERAS OF THE HEART

I have never been afraid to suffer a stroke. I am not typically one of those medical professionals who feels in their own bodies every illness they learn about during their studies. But now I sensed a twinge in my chest. It was not the sort of pain patients describe when they have a stroke. I did not feel a sort of wrenching, but rather a calling. Where did it come from, and where was it pulling me? The only thing I knew was that something in my life was no longer in tune; something had changed, but what?

From the outside, everything was fine. I loved my profession. I loved to operate. I loved my wife and my children, we lived in a nice house, we were all well. In my spare time I liked to sail, to cook and to read – even though I read less than I would have liked as I lacked the time. All in all my life was rich and fulfilled; I felt that was what I deserved, as I had worked hard for it. To get to my position, one had to serve for many years and put up with a lot, swallow a lot from the gods in white who wield the sceptre known as the scalpel. And you never know whether you will one day be admitted to that illustrious circle. There is no guarantee you will succeed. You also have to find out for yourself if you can cope with this profession and its responsibilities and if you have the hands for it. And you have to find a heart surgeon who will take you under their wing and believe in you.

The first five years, if you hold out, are spent in the realm of hope. As heart surgery requires a lot of personnel, there are scores of assistant doctors. You can calculate how long it will take if ten colleagues have their turn before you and if each of them takes five years ... it is not going to happen for you. But there is also a certain possibility in the long wait, as many lack the stamina – or should I say the capacity? – to suffer. They drop out. You need an iron will to continue – and it is easier the less you are in touch with yourself and your own needs.

Stamina will also recommend you to your superior, who at first won't even know your name. Then comes the day when you are allowed to stitch up a leg after veins have been extracted from it for a bypass. And you don't know if that will be it, or if one day you will be allowed a go at the holiest of the holy, the heart. It helps if you never contradict your boss. If you complete everything perfectly. If you never reject a weekend shift. And if the boss asks on Friday evening if you can stay a little longer, you don't say that you have been here for two days straight and today is your mother's birthday – you say: yes, of course.

At university clinics it is expected that you write research applications and academic papers in your spare time. That will enhance the department's reputation, and at some point, hopefully, the boss will notice that you are serious. Then you are a good assistant, no longer a nobody. Maybe you will even be greeted by name. And then you may be allowed to saw open a chest. It is a sign of trust if the surgeon lets you do that. And then close it. Or connect the heart-lung machine.

And so you continue to serve, and become tougher and more hardened. You swear to yourself that you will never become such an unjust god in white. But by the time you have conquered the throne, once the right to operate on hearts has been bestowed on you, who will you have become? Often simply what you hated so much before. Is this inhuman, or rather all too human?

The pressure on heart surgeons is enormous. In a recent study, primatologist Frans de Waal observed the behaviour of surgical teams in the operating theatre. It was mostly cooperative and friendly, but in some cases there were serious conflicts involving verbal violence – sporadically instruments even flew across the room. Particularly during long operations – so normal for heart surgery – stress increases dramatically and the probability of conflict rises. If there were predominantly or exclusively men in the theatre, the frequency of conflict was twice as high, compared with the figure for male and female teams. For de Waal the parallels with the apes he has spent his life studying must be obvious.[1] Doctors are only human after all.

How do I treat my assistants? I asked myself one day, and then immediately thought: *What sort of question is that? What had happened to me since I held the nipple piercer's hand? Was I contaminated with compassion? Heart surgeons are cool and never lose control. Never. They don't cry, not even secretly. Fearlessly they foil the plans of the gods. But what if the gods took revenge? What if the gods without a white coat retaliated and sent a bolt of lightning out of the defibrillator?* What strange thoughts I had! Was I headed for burnout? What was wrong with me? And what was wrong with my heart? This twitch – no, it was not a pain, it was a calling.

Where should I go? And how much time did I have? Timing is tight with a stroke, every minute counts. Severe pain in the chest is typical and can also be felt in the arms, back, stomach or neck. In men, anyway. In women, it can be different. They don't feel the pain so severely, and it can 'only' be a pain in the arms or jaw, perhaps accompanied by malaise, nausea, vomiting or shortness of breath. Their symptoms may therefore not be taken seriously: perhaps it's just an upset stomach? Precious time is lost.[2]

But there is also the 'silent stroke'. You feel nothing. Absolutely nothing. Not for a long time. It is often detected by accident. So it was with me when I held that young man's hand. This was my diagnosis-by-accident. This touch had blasted calcium deposits from my armoured heart. I heard different voices in my heart and brain which I could not reconcile; they were wandering through me like those deposits. And if I was unlucky, something would shoot into my brain, I would suffer a stroke ... or had it shot into the brain already? And what were these chimera ... these new chimerical sensations of my heart?

The stroke

During a stroke, an artery is acutely blocked, the heart muscle dies and arrhythmia occurs. There is often a life-threatening complication as well. Immediate emergency measures are necessary; a small metal tube is implanted to replace the blocked section. Artery calcification does not happen all of a sudden; it is a process called arteriosclerosis, for which there are risk factors. The best-known of these are high blood pressure, smoking, drinking, obesity, lack of exercise, and stress. The sediments in the arteries are often called calcification; doctors also call them plaque.

Many people survive a stroke, and I sometimes get the impression that it is treated like the trophy of a high performer: I have achieved the maximum, I played really hard, I had a stroke. A similar prize goes to the burnout patient. All others are low performers. It seems to me that the effects of the performance delusion have changed since my childhood. Previously the most pious people had apparitions, today the toughest burn out. Some professions seem predestined for a stroke, in the same way that surgeons always have back problems, as they stand at the operating table for so long – and had to bend down for a long time before being crowned surgeons.

From the moment the arteries in which the stroke manifests itself are blocked, tissue dies. It is not all dead immediately, it takes a few hours – at least on the anatomical heart.

Yes, of course on the anatomical heart, what other heart would there be?

The other one, a voice in me said.

Apparently something had indeed shot into my brain. Which other heart? Was I hearing voices?

Yes. The voices of your heart.

Hearts don't speak. Hearts are pumps.

What if the anatomical heart was important for life, and the other heart . . .

. . . which other heart . . .

. . . was important for the spirit.

Rubbish. The brain is responsible for that. Everything important happens in the brain, we know that. We feel in the brain, it is the seat of consciousness.

Well, in that case I could calm down. I didn't have a heart attack but a brain attack. Or a heart attack without a heart?

The choreography of surgery

A nurse knocked on my door and told me, 'Dr Friedl, your patient is being washed.'

'I'm coming,' I replied.

Before a heart valve operation, a patient's torso and legs are washed thoroughly with disinfectant. Why the legs? Because in rare cases it is necessary to lay a bypass. For this, a vein is taken from the leg, hence the legs are disinfected as well.

I left my office and walked towards the operating theatre. I approached the cathedral of the clinic, and the temperature seemed to drop a degree. I entered the door code.

Peep, peep, peep, peep.

Click.

I stepped into another world. It appeared to be under water. A submarine, maybe. The door closed behind me, and I forgot everything that had just been on my mind. My heart was silent – and I thought this a necessary prerequisite for being able to help another person's heart. Professionalism, performance, hard as steel and free of any emotion.

In the change room I stripped down to my underwear and put my clothes in my locker. The green scrubs were ready, my work attire. This was the ritual of putting on my heart surgeon's armour. I passed the doors with their portholes into operating theatres. I did not feel anything anymore. I was functioning well. That was important because my opponent was powerful. I was fighting with death. During the minutes before an operation I always feel very alone. There is no safety net for the patient. Will I do it well? Will it be successful? What complications might occur?

Now everything depended on me. The responsibility was weighing on my shoulders. I must never underestimate the danger, even though it had long ago become routine. Nobody must notice that I was afraid of failure. I concentrated on the images and video I had just been looking at in my office. The patient's heart from all angles, black-and-white and occasionally in colour, 3D, pumping capacity, blood flow, location of the aorta. I recalled my strategy. How exactly was I going to operate? Which difficulties may arise, and what did my countermeasures look like? Which assistant was going to support me, which nurse? Could I rely on both, how well did I know them, what could I expect from them, could they think for themselves or would they constantly need direction? A heart operation requires at least four hands. I could not do it alone. It takes a team to help a heart.

I looked through the porthole into my theatre and saw that everything was ready. The patient was on the table, washed and covered. I put on a surgical mask and magnifying glasses. I washed my hands and underarms with soap and disinfectant for five minutes. An infection during an operation will have terrible consequences. The patients lie in intensive care with an open sternum for months because their wounds won't heal. Thorough disinfection also means respect for and appreciation of the patient. Ignaz Semmelweis, pioneer of antiseptic in hospitals, was called crazy 150 years ago because of his hygiene instructions, which many deemed exaggerated. If they had taken his findings seriously, many people would have been spared a lot of suffering. But back then it was not yet known that bacteria cause illnesses.

Now I was ready. I opened the last door to the theatre with a foot-operated switch so as not to have to touch anything. It fell closed behind me without a sound. I greeted those present. The room fell silent upon my entry. I was known to prefer a quiet atmosphere. Other colleagues would constantly crack jokes or chat relentlessly. A colleague from Bavaria, a superb surgeon, liked to open hearts in a sort of Oktoberfest atmosphere, kept up until the sutures were completed. I can't stand anything like that. And as I am the surgeon, I set the tone. With me, it is quiet. I like a focused silence, interrupted only by the bleeps of the heart rate monitor and the blowing of the respirator, the clatter of instruments in the kidney basin, the holding of breath and deep exhaling when a difficult stage has been completed.

The nurse dressed me, holding up the gown for me to slip into. A second nurse stood behind me and passed me the gown's strings. I took them and turned in a choreography

I had danced thousands of times, tied the cords in front of me and finally slipped into the gloves that were being held out to me. After a few turns and hand movements I was sterilely dressed. My own heart was switched off so I could operate on the patient's heart. Did it have to be like that? Would I be able to operate if it were different – and how did that even work, switching one's heart on and off? How would this patient perceive it? Which half of humankind did he belong to? Did his self reside in the heart or in the brain?

The nurse handed me the bone saw. I switched it on.

THE COLOURFUL NEUROSHOW

During my search for the heart I kept coming back to the same questions – even though the origin of feelings and thoughts was considered fully explained. They came from the head; neuroscientists had proven that with their images reminiscent of modern paintings. And the head, the brain, is the boss – the control centre. Attached to the brain is a body that needs to be optimised – a kind of computer governed centrally by the brain. We have become used to this perspective. It is scientifically up-to-date, isn't it? The many esoteric publications which describe the heart as the centre of all-knowing love stand against it. These two ways of seeing human beings are in competition with each other.

Navigating between them are the readers, who often do not know if they *are* their body or *have* one; they don't know who is talking when they say 'I' – only the head, or the heart too? – or what the story is with the soul. It is not rare for the talking I or Self to experience itself as separate from the body, or else its chief supervisor.[1] 'I would like to do this or that, but my body keeps getting in the way.' In general, the misconception prevails that we are close to deciphering the brain's code; all the colourful images of brains, no longer restricted to academic publications, contribute to this impression. And who knows – maybe in the future people will proudly show each other their brain scans as well as the ultrasound images of their unborn children.

I found three sentences that summed up my thoughts in Ian McEwan's novel *Saturday*, which revolves around the life of a neurosurgeon: 'For all the recent advances, it's still not known how this well-protected one kilogram or so of cells actually encodes information, how it holds experiences, memories, dreams and intentions ... But even when [the brain's fundamental secrets are laid open], the wonder will remain, that mere wet stuff can make this bright inward cinema of thought, of sight and sound and touch ... Could it ever be explained, how matter becomes conscious?'

Neurology's imaging technologies have enabled us to take coloured photographic images of the brain. But do they really depict feelings, thoughts and motivations? Brain researchers enthusiastically explain 'Here!' and 'See there!', as if they had discovered banners displaying logical thought or love. In reality, colour in the brain is by no means direct evidence of nerve activity, a fact that is quite frequently omitted. Instead, it is direct evidence of blood with high oxygen content, which is delivered by the heart. This blood flow flashes up as colour in functional magnetic resonance imaging (fMRI), and from this it is concluded that the nerve cells in this part of the brain are firing at this moment. If we feel pain the 'pain centre' glows, and if we have strong feelings the centre of emotion does the same. This is a wonderful scientific method that has revealed a lot about the functions of the brain's components. However, it is often erroneously concluded from this glow that emotion or pain or whatever else 'originates' in this spot.

Neurophysiologist Ernst Pöppel alludes to the fact that some areas of the brain, for example the insular cortex, appear to have multiple functions – 'emotions, body awareness, concentration, sex, desire, sense of time or pain' – depending on which study you review. The colourful images mentioned above do not have the straightforward explanatory power many will have us believe. The method is fuzzy; the newspaper *Die Zeit* even talks of a 'big neuroshow' in an article. Now some researchers have suggested that inner experience is in fact created in several cross-linked brain locations simultaneously.[2] Psychiatrist Thomas Fuchs comments: 'It is completely unclear, however, how all these sub-functions are coordinated and integrated to form homogenous consciousness.'[3]

Does this consciousness originate solely in the brain? Fuchs is a proponent of embodiment. According to this theory, the whole body is the home of emotion and consciousness. For these to develop, a body and communication with the environment are required. Embodiment considers the whole person, not just the blood circulation in certain parts of the brain. The (whole) body acts as a medium of emotional perception through its resonance and mobility, and it is engaged in constant exchange with the environment.[4]

If you hit your foot on a doorstep, the pain is first felt in the toe. For you to be able to feel the pain inside you, you need to have a body with a heart, nerves and brain. Thus, pain does not originate in your brain, but your brain knows where it hurts. You can switch off such pain with a general anaesthetic. You can also apply a local anaesthetic to the toe or interrupt the transmission of pain at the spinal cord. Or repair the doorstep before your foot hits it again.

The brain cannot feel pain; it does not have any pain receptors. A headache is not proof of the brain's sensitivity to pain; rather, it is an irritation of the brain's 'wrapping', meaning its lining, which is equipped with pain receptors. The brain has to rely on our sensory organs to supply it with information about the everyday world. Some of these organs are very familiar to us and their signals immediately become conscious experience: hands, skin, eyes, ears, nose and sense of taste.

But we also have inner sensors, as for example we feel when we need to go to the toilet or need to eat when our blood sugar drops. Fuchs believes that 'the brain itself thinks nothing . . . It is always the whole person who perceives, ponders, decides something, remembers and so on, not a neuron or a cluster of molecules.'[5]

The initiators of the Human Brain Project see this differently. They argue that the brain governs our body and that conscious perception originates exclusively in the brain. In 2013 they began a project to build a model of the human brain in the shape of a supercomputer; it was funded to the tune of 1.2 billion euros. In 2015 the first results were rather disappointing: with the phrase 'Thick skulls, false promises', the *Süddeutsche Zeitung* newspaper addressed the project's lack of progress. Andreas Herz, Professor of Theoretical Neuroscience, even called it a 'sham'. IT algorithms could not, he said, be used to recreate properties of the actual organ; a 'one-to-one simulation of the brain is definitely not possible'.[6]

Traditional neuroscience views the brain as a kind of biocomputer and undoubtedly has been successful in examining its components. If a certain component falls ill or is destroyed, the affected person's behaviour or sensations change.

It is similar to changing a specific processor or destroying the motherboard in a computer or TV – the images change, the screen goes dark, the information processing no longer functions properly. That is undisputed. But you cannot conclude from this that the information is *created* there. These electronic components, as well as the components of our brain, are necessary links in a chain that helps create the image. The information, the code for the content of the image or the processing power, is located in the software. And if we go even further back, the information comes from the brain of the software's programmer. And how does the information get into the software developer's brain? Does it originate there? When we watch a movie, we know the actors are not sitting in the television set but were filmed; the film was then edited and now it is being broadcast. So to watch TV we need sender and receiver, coding and decoding, as well as a range of devices and components.

No one today is surprised by the fact that information 'flies' through the air and even space via wireless networks and mobile phones. If you transfer this principle of wireless communication to human beings – using terms such as 'transmission of thoughts', 'intuition', 'compassion' and 'belief' – we are halfway to esotericism and leave measurability, and thus scientific respectability, behind. There's an old saying: if it can't be measured, it doesn't exist. But this already seemed dubious to astrophysicist Sir Arthur Eddington 100 years ago. With a wink, he declared that humankind had always been obsessed with measuring and defining everything in the quest for knowledge. He used the example of catching fish with a net. Scientists will decide that fish-catching is defined as an activity where a net is being used. But you might ask: what about the small fish who slip through the net? The scientifically correct answer is: what I can't catch with my net is by definition not a fish.[7]

Not everyone makes it quite so easy for themselves, but I absolutely see such tendencies in today's world, which is highly fixated on measurability.

Are emotions and thoughts one and the same in the brain? No one would say: I think myself fearful. I think myself desperate. I think love. We *feel* fear, despair and love. It is an immediate physical experience if we feel queasy with anger, or if we have butterflies in our tummy or joy in our heart. 'We experience ourselves and the world not inside the skull but as incarnated, corporeal beings,' says Thomas Fuchs.[8] We experience them with our whole body. The light-show in the brain is, in any case, no proof of increased nerve activity, but direct evidence of increased blood flow with all of its information, coming directly from the heart. From the heart!

Where is our home?

A study of one hundred and twelve students, forty-two of them female, asked participants if they felt their Self was located in their heart or in their brain. Surprisingly indifferent to the findings of neuroscience, 52 per cent voted for the heart, versus 48 per cent for the brain. Interestingly, this study found that where someone locates the Self influences their personality, emotions, decision-making and performance. A 'heart person' reacts, decides and acts differently from a 'head person'.[9]

Brain and heart have particular importance in the history of humankind. Plato viewed the brain as the seat of the soul and the source of reason. His pupil Aristotle, on the other hand, had reason and soul emigrate to the heart. He saw the heart as the place from which warmth, blood and life originate; therefore it must also be the centre of the soul and human reason. Over the next 2000 years, this examination of the soul was extended with different concepts and ideas by famous personalities such as Saint Augustine, Leonardo da Vinci and René Descartes: the soul migrated from the brain to the heart and back again. All hypotheses remained unproven.[10]

Maybe the truth lies somewhere in the middle. I don't mean it lies in the throat, but rather that the soul is a union of heart and brain. Unfortunately we still know very little about the connections between the two organs. For centuries they have been played off against each other, depicted as opponents with different ambitions. We reflect this understanding when we believe our head advises us differently from our heart in decision-making. People think of themselves as 'heart types' or 'head types', as being driven by their emotions or their intellect. Both are established within us. They remind me of secret lovers. Romeo and Juliet inside us, intimately connected. If one organ is sick, the other will often fall ill as well. Many patients, for example, become depressed after a stroke, and illnesses of the heart valves can cause strokes in the brain. If the blood flow from the heart is seriously disturbed, personality changes and even dementia can occur.[11] Vice versa, permanent stress and resentment are sheer poison for the heart: its pumping function may decrease as a result. It is not merely a saying but a fact: we can be scared, or bored, to death. Overstimulation of the brain is transmitted to the heart and causes fatal arrhythmia.[12] So if it becomes unavoidable, the Romeo and Juliet of our body join each other in death. They are inextricably linked in cardiac death and brain death.

We all know that our heart also hurts when we feel great sorrow. In extreme cases such as deeply felt disasters – say, the death of a family member or friend – the heart may tighten and cramp. People feel tremendous pain, as with a stroke. But it isn't one. The cause is in the brain, which uses nerve tracts to send more stress signals to the heart than it can handle. When screened, it will look constricted, like an octopus that has been caught in a pot. This is why the first people to describe a heart thus afflicted used the poetic term 'takotsubo syndrome', *tako-tsubo* being the Japanese word for a fishing pot used to trap octopus. It became known widely, and memorably, as the 'broken heart syndrome'. The new disciplines of neurocardiology and cardiac psychology deal intensively with the relationship of heart and brain, also called the neurocardiac axis.[13]

As a heart surgeon, I was of course also interested in the question of where my colleagues from the field of brain surgery would locate the soul. They are, after all, doctors who have seen a brain and held it in their hands! Here, too, opinions are divided. For some, consciousness comes down to electrochemistry in the brain, and they cannot accept anything else. Well-known Harvard neurosurgeon Professor Edward R. Laws is one of those who think it possible that the body together with the heart is the seat of the soul. Others view the soul as immaterial, immortal and impartible.[14]

Science still owes us a definition. The study mentioned at the start of this section inquired about the location of the 'self' – and that is suspiciously close to the 'soul'. But the study's authors carefully avoided using the term 'soul', maybe because 'the self' sounds more scientific. The term 'soul' is overused in esotericism, religion and philosophy. So academics in the natural sciences and humanities fill new skins with old wines: they use terms such as 'the self' or 'the psyche'. The word 'mind' has recently been favoured internationally, even though that term can refer to a multitude of things somewhere between reason, soul, brain and spirit.[15]

In the study most participants located their 'self' clearly and definitively, and could not imagine deciding differently. The idea that men are generally more 'in the head' and women more 'in the heart' was confirmed, but only just: of the 52 per cent voting for the heart, 64 per cent were women. 'Brain types' described themselves as rational, logical and unemotional in their dealings with others, whereas 'heart types' found more friendly words for themselves: emotional and warm in interpersonal contact. The results remained largely similar 285 days later, so weren't due to a momentary mood.

Where would you place yourself in this study? Whatever your answer, you should take care that your heart and your brain are well linked. One is not worth more than the other. Only their connection makes you a whole, indivisible human being.

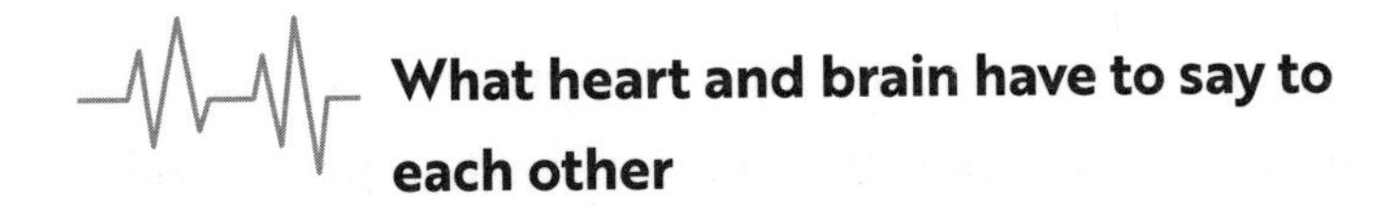

What heart and brain have to say to each other

International research teams have only just begun to decode the secret messages and understand the full extent of the intimate relationship between the two organs. Messages are endlessly exchanged within us via physical pulse waves, our circulatory system, our autonomic nervous system, hormones and neurotransmitters – and heart and brain influence each other even through their magnetic fields,[16] as they speak the same language, which has a rhythm, a beat. In this rhythmic activity of biological signals, information and messages are encoded – just as they are in modern telecommunication. When we talk to someone, we not only send out word signals but also incorporate body language. We find people more convincing when their words and gestures appear harmonious. A recent study found that people who put their hand on their heart when speaking were considered more believable. And they themselves found it much harder to lie when doing so.[17] Touch also plays a role in the connection between heart and brain. Not only does their language consist of biocodes, which we are still far from having decoded, but in a way the heart also embraces the brain – a phenomenon which is clearly visible.

Pulse waves

American neurosurgeon James R. Doty writes that when he opens a skull and looks inside, the whole brain pulses to the rhythm of the heart.[18] No fewer than four arteries run from the heart into the brain and continue there. The heart's pulse waves alter the shape of the elastic walls of the surrounding brain cells; this creates electric brainwaves which relay the messages from the heart. The young research field of biophysics demonstrates impressively how mechanical forces affect our cells and even our genes.[19] In the heart's pulsation, which is rocking and squeezing the brain mass, biophysical information is encoded that may influence our personality and emotions.[20] But that's not all. You may know that the brain swims in cerebral fluid. This is excellent packaging and protects the brain from injury. But it also gives it buoyant force, making it light. Otherwise the organ weighing 1.3 kilograms would lie rather heavily in our head. Floating like this, the brain weighs only fifty grams. For the fluid to remain as clear as a mountain stream, it is kept circulating by the pulse waves from the heart.[21] It is the job of a craniosacral therapist to sense the fluid's rhythmic pulsation and identify any blockages.

Hormones

A human being can only win a fight or solve a task if their heart and head act together. Sometimes it seems wise to run. Is this the sign of a cowardly heart? No, because especially if we run for our lives, we need a strong heart. And we need adrenaline – which comes not only from the adrenal glands but also directly from the heart. No matter if we fight or flee – adrenaline and its relative, noradrenaline, make us abruptly wide awake and increase the performance of not only the heart but also the brain.[22] 'Necessity is the mother of invention,' they say. Adrenaline and noradrenaline from brain and heart contribute to this. Some people yearn for this special kick, when our heart beats in our throat and our thoughts become fast as lightning. They call themselves adrenaline junkies and love danger. One person's joy is another's sorrow; they are as close to each other as heart and brain. They rejoice together and can both independently release dopamine (a happiness hormone) if the situation has been mastered, the fight won, the life saved, the stressful incident resolved, the adventure survived. And if life hits us between the eyes, dopamine will also help us to continue and get motivated again. Even though it may stumble during a crisis, our heart will not stop beating, and our thoughts will readjust after a while. Heart and brain go hand in hand, in good times and bad, and speak the same language of hormones and neurotransmitters.

Electromagnetic fields

The synchronised discharges of muscle and nerve cells generate dynamic electromagnetic fields. These have long been used in medicine and provide medical professionals with reliable indications about the state of heart and brain. The signals of an electrocardiogram (ECG) tell your doctor if there is, for example, a disturbance of the blood supply. Neurologists are more interested in the waves in the brain, which they analyse using electroencephalography (EEG). From this, they can tell if we are attentive, relaxedly dreaming or if cramps are causing an epileptic fit. If the electromagnetic fields of heart and brain can communicate so much to us medical professionals, they must also tell each other all sorts of things. That, however, is their well-kept secret. These messages are complex and have not been even partially decoded. We do not know how heart and brain attract, repel or influence each other via their magnetic fields. Experiments with cells allow us to conclude, however, that there would have to be a whole lot going on here in secret. Because when you release isolated cells in an artificially created electromagnetic field, this influences the permeability of the cell membranes and the cells' ability to communicate with the environment. It also affects our genetic make-up and the cells' ability to reproduce.[23] This, in turn, can have to do with love . . .

Nervous system

The heart even has its own 'small brain' with over forty thousand nerve cells. Or should I say the heart has its own head? Even though it is connected with every cell of the body, it can act quite autonomously and independently.[24] The number of nerve cells is decidedly smaller than in the cerebrum or the gut. But as we can see with many a fellow human being, the sheer mass or the mere existence of brain cells has not necessarily anything to do with intelligence or how to employ it. To use a quote attributed to Albert Einstein: 'Not everything that can be counted counts, and not everything that counts is countable'. The heart has a memory, and as far as we know today it acts in a time frame between milliseconds and minutes.[25] It is the organ for the present, its task is to create life in every second. Its first interest is what is happening at the moment, what is indispensable to life, what needs to be done now.

One thing, however, really astonished me. More nerve signals travel from heart to brain than the other way around – an unbelievable 80 per cent come from the heart.[26] The neuronal heart messages first arrive at the brain stem, the same as all other information which the autonomic nervous system has collected about our body and from the organs: via blood pressure, metabolic processes, body temperature and – very importantly – breathing. The brain stem houses the centre of our autonomic nervous system, whose name suggests that usually we cannot influence it. All inner body functions are permanently operated, adjusted and regulated there, millisecond to millisecond. The sensory organ that is the heart sends its messages to the brain via nerve post with high accuracy, and the brain minutely and subtly sends its information to the heart.

The result of these multifaceted connections between heart and brain, which are unknown to most people, is called heart rate variability.

Heart rate variability

Most people want a regular heartbeat, which becomes faster only when they hurry and slows down when they spend time with their loved ones. If you measure accurately, however, the heart also beats chaotically and irregularly when at peace – as discussed in *Circulation* and *Nature*, the most important journals for heart research and science respectively.[27] The interval between heartbeats, too, is by nature always different, never the same; it varies by up to several hundred milliseconds. So the healthy heart is by no means the Swiss clock it was long thought to be. A conventional ECG simply measures the heart rate inaccurately, taking the mean of the intervals between several contractions. Therefore intervals appear to be of equal length, and the heartbeat regular. So while this rough timeframe of the heartbeat as displayed by an ECG seems very stable, chaos reigns on the inside with multifractal, non-linear and spontaneous fluctuations.[28] Chaos, in this case, implies that we do not know what it means – it does not appear logical. Imagine you hear a language you don't understand. For you it is merely a collection of sounds and noises that seem chaotic and do not appear to make any sense.

In other words, the heart does what it wants, and it is absolutely unpredictable how yours or mine will beat in the next second. The clock of the heart is like the clock of life. Every person ticks differently. If you've ever had the feeling that the voice of your heart is chaotic, that impression was not wrong. Nothing is ever completely and continuously regular and symmetrical in a biological system. No river, no tree, nor your thoughts and emotions.

The more distinct your heartbeat's spontaneous adaptability, the less worn out and the more elastic your autonomic nervous system has remained. Heart and organs are still able to react flexibly. And that is the big secret of a long life. It has been unambiguously proven: the higher the heart rate variability, the lower the probability of illnesses and the higher the life expectancy.[29]

This also explains something which has played a big role for thousands of years in traditional medicine: the subtle diagnostics of the twenty-eight types of pulse which have been assigned to different illnesses. The heart, the body's centre, is connected with everything, and sooner or later most illnesses affect our heart and circulation. And the other way around. Good doctors once had refined their sensitivity to a point which allowed them to recognise with their senses things which today only computers can find.

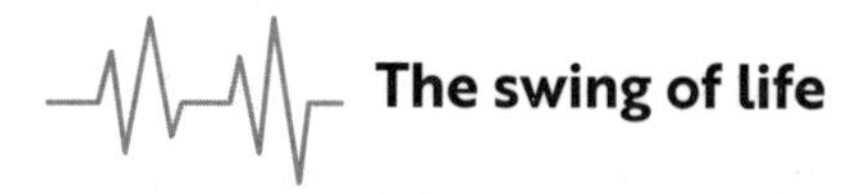

The swing of life

Organic systems are in a permanently flowing condition of simultaneous tension and relaxation.

We call such connections oscillating, swinging or trembling. Breathing, the hormone system, sleeping and waking are also oscillating systems. Even walking, our step, follows an ancient human oscillating rhythm. Everyone who likes to hike or go for walks will perhaps have experienced this: at some point you move according to 'your' tempo. You arrive at yourself. Walking is not a constant. You adjust to your environment, to the weather and maybe your thoughts. Or do you remember how you were on a swing as a child? My daughter sometimes sat on the swing under her tree house for hours, mused, sang, swang a little, jiggled, turned. And every mood, every thought, every idea led to a muscle movement which caused new swinging. How I would have loved to have heard her heart while she did that!

Even our genetic material, our DNA, swings. The new research field called epigenetics shows us very impressively not only how our genes define our body, but also how our behaviour defines our genes. Biochemist Erwin Chargaff, co-discoverer of our genetic material's secrets, says: 'A balance that does not tremble cannot weigh. A man who does not tremble cannot live.'[30]

The heart likes to oscillate or tremble – or one could say dance – in sync with the brain, and readily takes in its impulses; constraining and stimulating neuronal influences between heart and brain are an important cause of heart rate variability.[31] This dance is not a choreographed foxtrot or waltz in which one partner leads while the other follows. It is a subtle giving and taking, a leading and being led. It is impossible to predict how this dance will develop in the next second. It is the dance of life: at times subtle, at others wild, at times exhausting, at others very intimate. Life, too, oscillates and is not a straight line from the cradle to the grave.

A relaxed artist of life plays a special role in this dance. It is the vagus nerve, the tenth cranial nerve. It is regarded as the most important agent of our body's department for wellness and relaxation, also known as the parasympathetic nervous system. Old anatomists called it a stray and vagabond as it drifts through the body seemingly aimlessly, pausing a little here and there to have a chat with the organs. Its impulses connect all organs, and the more the vagus talks with the heart and the more the heart listens to it, the higher the heart rate variability will become. The heart can be seduced into relaxation within half a second.

Things are very different, however, when the sympathetic nervous system sends stress signals which may relate to work and performance. This nervous system constantly brings the latest news from the brain and reports crises and dangers. The heart, with its typical prudence, does not respond immediately but rather waits up to four seconds to see if the danger is really present or if it merely appears to be. If it were to permanently react to every warning of danger, it would quickly exhaust itself. It cannot afford that as it wants to keep dancing. Thus it sends its impulses back to travel to the brain via the vagus.[32] And they will dance happily ever after.

HEART TONE

Most people I knew had already been to New York, often while studying. But I had worked for some years as a senior doctor before I went there. When I did get to know the city, I visited the usual sights and a few other gems, such as the Rubin Museum of Art, where spiritual art from all over the world is exhibited. I had been meditating regularly for some time and was interested in a spiritual path of truth, independent from any particular religious or philosophical worldview. In my opinion there could only be one truth (there was, after all, only one heart), even though many paths may lead to it – just as there are many pathways to the heart, or roads to Rome. In the museum's foyer I browsed through an illustrated book and discovered a sentence by the Persian mystic and poet Hafiz, born around 1320. My eyes caught on two lines that gave me goosebumps: 'I am a hole in a flute that the Christ's breath moves through.'[1]

Suddenly I heard something. I did not know what it was: a sound, a tone being repeated. I had to pause, to listen; I was touched in my innermost being. It was a kind of peaceful humming, beautiful, warm, round, harmonious, like a warm oil bath for the ears. I opened them wide, listened intently and followed the call up a spiral staircase to the heart of the museum. Thus, I reached an exhibition entitled 'The World Is Sound', in the centre of which you could hear the syllable OM sung by people from all over the world, including several artists. A choir of light and sound. Buddhism and Hinduism regard the syllable OM as a spiritual force, as holy. OM is a metaphor for the most fundamental attributes of our existence, of its wavelike nature.

It seemed to me that both my heart and my brain responded to the OM, to this call – they went into resonance. I sat down and remained there until the syllable OM had penetrated deep inside me, until I felt that every one of my cells had absorbed this tone. I listened not only with my ears but with my whole body. The swoosh of my thoughts receded, and clarity set in. At the beginning was the word. Isn't OM a word? Yes, there could only be one truth, no matter if it was from the Bible or from Tantric writings.

Feeling fulfilled and blessed, I left the museum. At night in my hotel I realised that I was still feeling the tone under my skin. Yet no ECG could make it visible, no stethoscope could make it audible.

Heart mathematics

For decades, the Californian HeartMath Institute has been investigating the connections between heart and brain. It is called HeartMath because the researchers involved want to work out and measure the secrets of the heart in a scientific manner. For this purpose, they have developed a biofeedback device which makes one's heart rate variability visible on a laptop screen – via an ear clip with a cable and a USB connection. Different studies have shown that negative emotions such as fear, sadness, anger and worry disturb the communication between heart and brain.[2] An irregular stress wave will appear; even just seeing it makes you nervous. And we feel as awful as it looks when we have negative emotions – our inner self loses its beat. However, it is possible to escape from this condition and get heart and brain back into a balanced and relaxed connection. This state is called coherence, or you could call it consonance. The autonomic nervous system, breathing, brain and heart synchronise themselves and oscillate in a harmonious way.

Practising this is actually quite easy: it requires conscious natural breathing, conscious presence of heart and emotions and an alert spirit. In this way, your heart rate will become a little quicker with every inhalation and a little slower with every exhalation. The stress waves will disappear from the screen, and a harmonious, sinusoidal, coherent curve will replace it. Coherence is a natural state for babies and small children; their autonomic nervous system is still very elastic, their heart sensitive.

With HeartMath's method of changing your heart rhythm pattern you can return to this natural flow and reach a coherent (meaning harmonious and stress-free) connection between oneself and one's heart – something we otherwise strive to achieve using Eastern methods of relaxation such as yoga and meditation. While we can calm our brain only with difficulty (because usually we are constantly thinking something), we are easily able to pacify our heartbeat without stopping it – with a few calm breaths. And suddenly the heart will beat more relaxedly, and our pulse will cradle our brain, which is snug in its shell.

Before my first self-experiment with the biofeedback device, some weeks before I travelled to New York, I did not have the slightest doubt that my heart would present itself in a coherent curve. I had been practising meditation for a while, after all. So I would not have to prepare myself by thinking beautiful thoughts or breathe consciously. I would plug myself in and immediately see the well-formed waves that would show me everything was perfectly fine with me. Oh well, maybe I would meditate a little so my waves would appear especially rounded on the screen. That was the plan, at least. I connected myself to the device, closed my eyes and meditated for twenty minutes. I tried very hard to relax.

When I opened my eyes again, I expected to see a top performance on the screen. I was used to nothing less from myself. Always in the lead, always performing at my peak. But not this time. Even a layperson would have been able to recognise at a glance that I was stressed, given the hectic, irregular amplitude. The coherence I had expected looked more like the fever curve of someone critically ill. What a defeat for the meditating heart surgeon!

Such a silly device. I should have saved the money. I should have known that something like this was worthless. How could a mere apparatus recognise the value of my meditation! I was positively offended and put the thing away somewhere. And forgot about it. For a time.

When the heart sings

Two days after my light-bulb OM moment at the museum, I packed my suitcase as I intended to fly back to Germany the next morning. Suddenly I had a cable in my hand. At its end dangled – at first I didn't even know what it was – the 'silly device'. The malfunctioning relaxation machine. How had it gotten into the suitcase? I was about to put it away in a side pocket when I hesitated.

I was a different person now. I carried the tone in me. If the truth resides in the OM, my heart should also react to it. My performance should have changed. Should I give the device another chance? Or myself? I connected with the device and immediately looked at the curve this time; it hadn't changed since the last attempt. Again it reminded me of the fever curve of a critically ill person. But I was no longer like that, I was on the road to convalescence; I had begun to search for my heart. My heart! But would I ever be able to grasp it with my mind? To be honest, I did not really believe it anymore. I had to admit it would not work with the mind alone. I watched the screen and the jagged curve for a long time. I was no longer annoyed. It was simply there, and I accepted it as it was. I took a deep breath and then continued to breathe consciously.

At the start it was very low. Hardly audible. With every exhalation, a tone rose from within me. A very small and tender OM. It sounded fragile. A shy sound of the soul that expressed everything I felt at that moment. I gave in, no longer wanting to meditate my heart rate into a specific curve. A sadness and an age-old pain manifested themselves in this tone. But also the freedom to be able to let go. With every exhalation I let go a little more, and the tone of OM grew from within itself. I would not say I made it, let alone sang it. It was simply there.

And then the miracle happened. The heart curve changed before my eyes. Which I could not believe at first. But it was not a hallucination. As if by magic, the nervous scribbles became a small wave. And then another. My heart made many little waves. My heart relaxed. And I with it. Perhaps for the first time in decades. It was so immensely comforting to observe that. It was so touching to see the harmonious curve of my heartbeat on the monitor.

All my thoughts vanished. I only listened to my voice and my breathing. The tone changed with the deep sensations. First there was gratitude and joy, and after a while it changed to self-compassion. When had I last felt compassion for myself? I could not remember. I had managed my whole life like a bureaucrat. At some stage all sensations gave way to a deep peace. I was in resonance with myself and the world. My heart was talking to me, and I answered with OM. No one heard me. No one saw me. I was alone and a long way from home. But maybe I had never before been so much *at* home as within this sound of the world, as the holy syllable is also called. Because existence of any kind is sound. Physicists have long proven that the whole universe is based on vibrations. And the curve of the connection vibrated on the screen, harmonious and round – coherent. The theory of embodiment was no longer just a concept but a reality I experienced.

WISDOM FROM THE HEART

I watched my colleagues in heart surgery. Did they have such heretical thoughts too? Was I still one of them, or had I diverted from the pure faith in the pump? Were we allowed to talk about it? How would they react if they knew I had chanted 'OM' in a New York hotel room? Or if they knew I now met with people who examined their innermost heart without using a scalpel or spilling a drop of blood? To gain clarity, I had to ask them. My first port of call was a surgical nurse I trusted, whose heart I thought was in the right place.

'Tell me,' I asked her in our kitchenette, 'one feels love in the heart, right?'

She stirred sugar into her cup without answering. Maybe my question was too personal?

I asked others, noticing only later that they were all women. Did I assume women would have greater competence in matters of the heart? Did I fall for the stereotype that women have hearts while men have brains? At times I felt as if the convolutions of my brain were in a knot – it was just so difficult to distinguish the anatomical heart from the one that feels and acts wisely. On top of that, I was in a terminological muddle. Did I mean the soul? Why were all these essential parts of being human so vague and blurry, and not just in everyday language: philosophy, too, had not found an answer, and psychologists used the terms 'soul', 'self' and 'emotions' interchangeably, depending on the school of thought they followed. As a scientist, I would have preferred clarity. A big unknown, an unambiguous formula, a result, done. Why did I even want to know it so precisely? Wasn't it enough to know that the voice of the heart exists in principle? To connect myself to the biofeedback device now and then to check my curve? No, that was not enough. I sensed there was more. After a cardio technician and an intern, both female, had confirmed that the brain was responsible for feelings – 'Well, Reinhard, that's common knowledge these days' – I tried a male assistant doctor. 'The heart is the source of consciousness, isn't it? Without the heart there wouldn't be any thinking.'

Wide-eyed, he stammered: 'Yes, of course.' But I noticed that he seemed to think I was mildly insane. (He, in turn, was certainly not – or else he would have contradicted his boss.)

If I continued like this, they would suspect I was not quite all there – but were heart surgeons ever all there?

'Can I hear the joke too?' asked the anaesthetist who sometimes brought in wasabi nuts hot enough to numb your tongue. Apparently I had been chuckling to myself, alone in my office with the door open.

'Love dwells in the heart,' I said.

'Where else,' she said, shrugging, and walked off.

I laughed. And noticed that my heart had become a lot softer. More flexible, so to speak, and that meant healthier, too.

A flexible, elastic heart also means more resilience, the ability to master crises, to be tougher. I badly needed that attribute right then. Because as a heart surgeon, one could say I was skating on thin ice. But I was well prepared: various studies had shown that some of the neuronal messages of the heart travel through the gate to consciousness, a region of the brain called the thalamus. There they directly influence the function of higher brain centres and our manifold inner experiences. It has been found that people with a flexible heart are more emotionally balanced when tackling difficult tasks. They also have better self-control and are less easily frustrated. Furthermore, people with a high heart rate variability can remember questions and problems better and have better short-term memory.[1]

I wasn't really worried, though, that I would forget my search for the heart – since it was also engrained in my long-term memory.

Besides all this, an elastic heart also lifts one's spirit and subjective sense of wellbeing, and enhances our social behaviour.[2] This variability of the heartbeat was the hot lead in my quest to prove that the heart's physical voice was inextricably linked with our emotions and thoughts.

The elastic heart even reveals something about our capacity to empathise: the more a heart can adapt, the more compassion a person feels.[3] This can also be useful for one's self-care: for example, if you like to eat and raid the fridge at night. A recent academic study proves that people with higher heart rate variability are better able to withstand hunger pangs and reach their dietary goals.[4] But does that prove the wisdom of the heart? What even is wisdom?

A wise person considers the wellbeing of others, not just personal advantage; they consider the opinions of others, and accept other viewpoints; and they consider the possibility that they don't know everything, so they seek advice and are able to make compromises. We see the opposite of such wisdom in current world politics. Intelligence alone is by no means enough to make wise decisions. History is full of 'brainiacs' who often acted in a selfish and discriminatory fashion. In 2016 a scientific study proved that egotistical points of view stand in the way of wise decision-making and separate us from the signals rising from the heart.[5] One hundred and fifty healthy participants were asked to find answers and solutions for problems in areas they were familiar with:

health care, education, environmentalism, politics, taxes and security. They were asked to take either an egotistical point of view or one where they removed themselves from the equation. Simultaneously, their hearts' ability to adapt was measured. The higher their heart rate variability, the more prominent was their ability to make wise decisions – but only if they adopted an inner perspective in their brains that considered other people's interests. If, however, they maintained an egotistical point of view, the heart could not do anything and higher heart rate variability could not exert any influence on the wisdom of their decisions. We can only act wisely when an open, adapting heart is connected with an open, sophisticated mind.

Heartbeat and brainwave

I had always believed that heartbeat and brainwaves did not have a lot to do with each other. Now I learned that the heart can also make a wave which influences consciousness. The wise heart is not merely a metaphor.

Very recently, the immense importance of breathing to the wellbeing of our physical heart was revealed. In a study of over 900 patients observed for five years, the survival rate after a stroke was significantly improved by changing heart rate variability through breathing (which is exactly what happens when you chant 'OM'). The data can even be consulted to estimate how long patients will survive.[6]

Most studies about mortality acknowledge risk factors such as smoking, diabetes, obesity, lack of exercise, elevated levels of blood lipids and high blood pressure. Today we know that reduced heart rate variability is also an independent risk factor – even one of the most important ones. Frequently it occurs at the beginning, long before an ECG shows anything or a patient's blood pressure goes 'through the roof'. So you have to detect it early. A recent study with more than 20,000 participants observed for between three and a half and fifteen years demonstrates extremely impressively that low heart rate variability increases the risk of cardiovascular disease by 32 to 45 per cent.[7] And at the very end of a long illness, you will have a patient's heart that beats barely at all, and inflexibly, the heart rate variability close to zero.

The inner world of those affected becomes poorer and unbalanced. They become prone to stress, hostility, bullying, depression and anxiety. The stress system, which is regulated sympathetically, becomes hyperactive. If this state continues for a long time, our body's energy requirements become excessively high, and we use more energy than we have at our disposal. At some stage this overdrive (or should I say overkill?) will lead to early ageing, illness and increased mortality.[8] But how can an unspecific trigger such as emotional stress make a heart sick? What are the underlying mechanisms?

Researchers at the renowned Massachusetts General Hospital and the Harvard Medical School proved in 2017 that for human beings the amygdala plays a role in this regard. It is the 'wild animal' in the zoo of our emotions, the part of our brain that becomes active when we face danger, fear or anger. If the amygdala serves continuously and unceasingly as a command post, our bone marrow will produce more and more inflamed cells. This will eventually lead to vascular inflammation and later arteriosclerosis and vascular obliteration. Two hundred and ninety-three people who had never suffered from cardiovascular disease were studied; they were observed for three to seven years. During this time, twenty-two of them suffered strokes, angina pectora, cardiac insufficiency, peripheral vasoconstriction or apoplexy. Why those twenty-two and not the others? What they all had in common was that PET-CT images had shown their amygdalas to be highly active. It was proven for the first time, and published in one of the world's most prestigious medical journals, *The Lancet*, that if our centre for anger, stress and fear in the brain is continuously activated it inflames the whole body and makes heart and blood vessels sick.[9]

The study's authors point out explicitly that so-called stress management (especially the kind that considers our feelings) is essential for the survival of our heart. Antidepressants and tranquilisers are not what is meant here, as their long-term effects are disastrous. Their work is based on blocking different receptors, and patients thus can no longer feel what should be felt. They become deaf inside. It is understandable for patients to wish for such drugs when stress and emotional pain become unbearable. You simply want it to stop. In the short term, tranquillisers may offer a solution – but in the long run the side effects are too serious. A study with over 15,000 patients receiving therapy with antidepressants has demonstrated that in the long term such medication leads to an increased heart rate and further reduction of the heart's variability.[10] Of course, alcohol, drugs and work (being a workaholic) are not solutions either.

So what can we do? The good news is: a lot, as nothing is carved in stone in our biological body – many processes are reversible. A healthy diet, quitting smoking, physical activity, losing weight, and consuming less alcohol work wonders. And there is yet another remedy: breathing.

Research into stress has long searched for something like a 'jammed gas pedal': a reason why we are often stuck in a hyped-up state. Today we know that usually the problem is not the gas pedal but rather a poorly functioning brake. In other words, the lack of relaxation.[11] It is the vagus nerve that possesses this brake function; it can relax us, can cool down a hot heart and increase our heart rate variability. But many patients are simply no longer able to find the brake pedal. Take the case of Mr Brown . . .

The enlightenment

Mr Brown came to me to 'have his pump checked out'. He was forty-nine, tall, overweight, and I did not have to ask his profession to know that I was dealing with an executive.

'I'm on several boards,' he informed me without my prompting. 'It's stressful. And I'm approaching the dangerous age now, right?' He didn't give me time to respond but continued. 'My blood pressure is too high, and I might as well tell you straight: I smoke. Can't stop because of the stress. And as you can see, I'm not exactly slim. But I feel good. Yes, good. But I'm not sure if everything's okay. Sometimes I feel funny. And last week someone at the firm conked out.' He clapped his hands. 'Bang. Gone.' He exhaled noisily and looked at his watch, as if wanting to know how much time he had left. 'I should exercise. But when?'

I opened my mouth to ask him a question, but he promptly shot me another one: 'How long does the appointment with you take?'

I could not answer that, as I was not yet clear why he had come to me. If the diagnosis was just 'strange', it could have been given by a family doctor, psychiatrist, psychologist or nutritionist.

I pondered how long I could stretch the time with him without stressing him out further, and replied: 'Half an hour.'

He sighed deeply. 'Okay. Half an hour, okay.'

'I would suggest—' I began, but he cut me off. 'Should I maybe have a blood test?'

'Yes, we'll do that too,' I said, although I had something else on my mind first. Patients like Mr Brown often want something done straightaway so they feel well looked after. Sometimes I hook them up to the ECG machine to calm them down. With Mr Brown, I wanted to start by listening to his heart. A little grudgingly he took his shirt off, presumably offended that I did not employ more technology for his health.

'How long will the test results take?' he asked.

'If you can't wait around for them, you can call later to get them,' I said.

'Good, good,' he nodded. He relaxed because I had heard his message that he was a busy man without much time. I, on the other hand, was acting as a sort of messenger for his heart, and what I heard when I put the stethoscope to his chest was worrying. Even more alarming was Mr Brown's blood pressure. 185/110 mmHg.

'I told you so,' my impatient patient said. 'It's always like that when I see a doctor. I have white-coat syndrome.' He tried to laugh. I sensed his fear and started to be a little afraid for him myself. He had a high-blood-pressure crisis.

'Maybe I will come back tomorrow,' he thought out loud.

'I can't let you go like this,' I said calmly. 'We'll do an ECG.'

He did not object. I had already heard that his heart was too fast, and the rate of 105 beats per minute confirmed this. Fortunately I did not find any signs of a stroke, as I had feared a little that I would. The lab results would confirm this later. I gave him an injection with emergency medication for his far-too-high blood pressure and asked him to remain lying down. 'I will have to make a phone call first.'

'I would like to take a closer look at your heart,' I suggested as an alternative, 'and you can also see it on the screen for patients.'

'Oh really? Okay, then. That's interesting. I'm rather curious.'

Together we looked at his heart in the ultrasound – echocardiography, as it is called. At first glance everything worked well, and the valves seemed okay too. They opened and closed impeccably. The walls of the left ventricle were thickened, as is common in patients who have had high blood pressure for years. But somehow the whole heart movement was not 'round' – the heart looked as if it were wobbling. Further testing confirmed my suspicion that the left ventricle's walls were thickened. This heart did not have a problem with exertion and contraction – it was the relaxation that was disturbed.

After half an hour, his systolic blood pressure had gone down to 140 mmHg.

'So everything's fine again,' said Mr Brown.

'You can't leave such high blood pressure untreated,' I countered. 'It leads to arteriosclerosis, damages the veins, and the risk of a stroke or a heart attack is high.'

'And what does that mean?'

'That the worry which brought you here is justified. And that you listened to the voice of your heart.'

Now he seemed pensive.

I took a deep breath and then launched into a somewhat different doctor's talk. 'You actually should start taking medication straightaway to reduce the high blood pressure and fast heart rate.'

Mr Brown interrupted me. 'I won't take beta blockers. A friend of mine's taking them, and since he started,' he pointed to his groin, 'nothing happens.'

'And how about you?' I inquired.

Silence. Finally he cleared his throat. 'They've got drugs for that nowadays.'

'Viagra?'

He nodded.

'And?'

'Terrific. But . . . Viagra doesn't help my back.'

'Your back?'

'Yes, I often have horrific pain in my lower back. Kills the sex drive. Apart from that I sleep badly.'

I was glad to have asked, because after these confessions it was clear to me that Mr Brown's autonomic nervous system was considerably defective. Contraction and exertion were dominant here: in the heart, in the head and in the back. But not where he most wished for it, between the legs.

Normally he would be prescribed medication to lower his blood pressure and a sedative or even a mild antidepressant. His risk of heart attack or stroke was very high. But I wanted to try another approach.

I explained heart rate variability to my patient and explained it was possible to influence it without medication, through breathing and relaxation.

'I'm not the yoga type.'

'Breathing,' I smiled. 'Not gymnastics.'

He sighed. 'Okay.'

I attached the electrodes to measure his heart rate variability. He became impatient again. 'How long will this take?'

'Twenty minutes.'

I was not surprised to see that the screen showed a stress curve. I asked Mr Brown to close his eyes and breathe. He wasn't able to keep his eyes closed, nor to breathe calmly. I liked him.

'In that case, let's breathe together,' I suggested.

So I sat beside him and breathed with him. Gradually his breathing became slower, his exhalations longer. He was following my lead. I asked him not to say anything, to trust me and just play along. Then I started with a soft 'OM'. After a few seconds Mr Brown joined in, which I was surprised as well as pleased about. But he still couldn't drop his jokes.

'Doctor, are we becoming enlightened now?'

Not as quickly as that, I thought – except maybe in MRI. A recent study examined what happens in the brain when people hear OM. Scientifically, the result was put like this: the blood flow increased in the 'left – dorsolateral – middle frontal gyrus, and in the 'right supramarginal gyrus'. Meaning the brain started to light up like a Christmas tree in regions that had to do with empathy and compassion.[12] That old acquaintance and relaxation artist the vagus nerve is also involved. Its signals allow the vibrations in the heart to increase, which in turn considerably improves the blood flow in the brain.[13]

If we had done an MRI scan of Mr Brown's brain right now, it may well have shown 'enlightenment'.

It is interesting that so far no one has examined how immediate, curative and powerful the chanting and breathing of OM is for our heart. To the best of my knowledge, having searched all relevant databases, this book contains the first reports anywhere in the world about the effects of OM on the heart.

When Mr Brown opened his eyes twenty minutes later and saw his coherent heart curve, he cried: 'Terrific! Can I buy this device somewhere? I would connect it every day.'

'It's not about the device. When you next connect it, such a coherent curve may elude you.'

'OM,' Mr Brown smiled. How relaxed his face looked now. Not one bit like the stressed executive's mask at the beginning.

I wanted to motivate him a little and explained: 'For me, this visualisation is a good first step to get in touch with my heart. The OM helps me control my thoughts. It is as if they fly away with every exhalation. And as I am curious, I have tried out many different ways of breathing. None was as effective as OM. The HeartMath device has shown me something I otherwise would perhaps not have believed – how much my feelings, thoughts and breathing influence each other and how much the heart is able to express all this. The device has also supported me. But it cannot produce any amazing feelings or create a state of deep meditation. It is all about your willingness to relax, to become involved with your heart.'

'Oh, health is so arduous,' he moaned.

'Illness is more arduous,' I replied.

He was silent. He appeared thoughtful, and considerably calmer. His heart rate had dropped to 80 beats per minute.

If before his appointment he had made any sort of decision already about what to do, he would potentially decide differently now, maybe even wisely. His heart and brain were now joined in relaxation. This would have tangible long-term implications for his health and life expectancy. I am convinced that his risk of a heart attack would drop, and his blood pressure would return permanently to normal. And I have solid scientific reasons for this.

'So I'm to de-stress and breathe?' Mr Brown assured himself.

'Among other things.'

'And sing?'

'Can't hurt,' I smiled.

'Do I get a prescription?'

'Yes. I also recommend a dose of conscious breathing at least three times daily before meals.'

And I would like to recommend this to you, too. As best you can, try to get in touch with your innermost heart. Perhaps you can hear your heart a little? It has been with you for so long, longer than you suspect, longer than you can remember. It was already beating three weeks after you were conceived; it is the first organ to develop. Once upon a time, you were only a heart.

HEARTS IN SYNC

Not that long ago I was on an emergency rotation to a small town in Mecklenburg. The emergency coordination centre told us the patient was thirty-four weeks pregnant. A premature birth was imminent. In big cities there are emergency doctors who specialise in newborns, trained specifically to look after premature babies. I, the 'regular' emergency doctor on duty that night, usually treated adults. My driver reckoned: 'Probably a false alarm. High-risk pregnancies are really well monitored these days.'

That was what I was hoping too, but to be on the safe side I asked the coordination centre for the nearest hospital with a gynaecology department. It was thirty kilometres away.

It was an early May evening, mild for this coastal region. Lots of people were having barbecues and the Saturday night was electrified with anticipation of the upcoming

Champions League final. With flashing lights we approached the patient, speeding through a concrete jungle. We turned at a garden allotment area covered in barbecue smoke, another concrete jungle, then we parked next to the ambulance that had arrived a few minutes earlier. Our arrival had not gone unnoticed: at three windows faces were visible, hair rollers around one, toplessness below another. On the ground floor there was a delicious smell of freshly cooked garlic, one floor up of coffee; an apartment door was open on the second floor – we were in the right place, as the name beside the doorbell told me. In a small bedroom, two paramedics were already looking after the patient, a blonde woman in her mid-twenties, her face red, her fringe damp on her forehead. *Why was she covered up to her neck in a duvet?* I wondered. She smiled at me, I smiled back. A paramedic was kneeling on the right side of the bed, taking her blood pressure. On the left, an oxygen saturation monitor was attached to her finger. My driver was preparing the ECG. An intravenous drip had been put in place already, and the glucose test was showing normal results. Everyone was busy, which in turn was making them feel okay. Don't just stand around. Help. Work through the medical to-do list. Yet the three men in their steel-capped boots, equipped for a crash on the highway, seemed out of place in this small bedroom with its photo wallpaper (did such a thing really still exist?). And where was the tummy that held my second patient? The mother-to-be hid her body from sight, the duvet up to her chin. She felt good, judging from the fact she was beaming. But it was best to ask.

'How are you feeling?' I inquired.

The smile became wider still. 'Good.'

'Are you having contractions?'

'Yes.'

'How often?'

'Every few minutes.'

'Then we still have time to drive you to the clinic,' I said with relief.

'I don't think so,' she replied, suddenly tense, and pulled back the duvet. I could not believe what I saw. The baby's head was at the pelvic floor. I could already see a bit of fluffy hair. The mother was so relaxed I had not expected the birth to be so far advanced. I had to take a deep breath.

'I'll quickly wash my hands,' I heard myself say. I had to think for a second. A premature birth in the thirty-fourth week. I had operated on the hearts of a few premature babies, some of them weighing less than 500 grams. But a birth was something entirely different. Besides, we weren't in a birthing suite full of neonatologists and paediatric nurses, but many miles away from a hospital.

On my way to the small, windowless and rather stuffy bathroom, it occurred to me that we lacked someone to boil towels. That's what they used to do in the old Westerns I had watched at my grandma's. That was the only thing that came to mind in that moment. Apart from the birth of my own children, this was the first birth I would attend as a doctor. But this was also a matter of life and death, and I was familiar with those. This little human being was about to barge into the world about six weeks too early, and I did not want to think about the horrible complications that might ensue. When I returned to the bedroom, the patient announced: 'It's starting again.'

A paramedic gave me gloves. I put them on and stepped closer with outstretched hands, to protect the mother's perineum and to be able to slowly 'unwind' the child, should that become necessary. Then she made a sound like suppressed sneezing three or four times. I got another surprise. The child shot out like a cannonball – straight into my hands.

My heart nearly stopped. The child was blue and limp. As quiet as a mouse. What now? You can work as an emergency doctor for twenty years without ever experiencing a birth. In my case, I had extra qualifications in the resuscitation of children. But in that moment I was numb. Helplessly I held a tiny newborn human being in my hands. At least my doctor's reflexes worked. I heard myself ask for a suction drain to suck away the mucus from the baby's mouth. During my medical course I had worked in a gynaecology department in Bangkok for a few weeks, where it was like a production line for children, around the clock, thirty beds in a maternity ward, separated by curtains. Such thoughts raced through my head at lightning speed – while I was highly alarmed as the little one was not showing any signs of breathing; the baby urgently needed oxygen. A paramedic gave me the suction catheter. My shock-frozen brain thawed. The little one had to breathe. If the baby would not breathe, its heart would soon stand still. Was it beating at all? My hands remembered my own newborn children. I turned the baby over, noticed for the first time that it was a boy, and then – there was no sound on earth more beautiful – he began to scream. He screamed powerfully and, it seemed to me, with a lot of outrage. I could understand that – there you were just seconds ago, suspecting nothing evil, in a nicely darkened environment, with chilled-out heartbeat music, floating in pleasantly warm amniotic fluid. Then you are pressed through a tight tube, deprived of oxygen, and eventually plopped out into a bright, cold and hard new world. He screamed his pain for the world to hear, and this primal scream saved his life. All life is suffering, Buddhists say. In this case, though, pain had led to life.

'Jakob,' the mother said.

'*Ja*,' I replied. Yes.

'Ja-kob,' she said again.

'Yes. Jakob.' Lovely name. The *Ja*/Yes at the start. And what a start it had been. Little Jakob was screaming like a furious animal now. His whole body was screaming, and his heart supported him. I no longer needed the suction drain. Jakob's lungs were working better and better, which had direct consequences for his heart. As the pressure in his thorax suddenly changed, his heart, too, adjusted from its state before birth to what was required now. The right side of the heart was pumping blood into the lungs. In mum's tummy the lungs had been bypassed as oxygen was supplied by the mother through the umbilical cord. The new blood paths from heart to lung are opened with the first breaths, and all bypasses closed – a fundamental change for the cardiovascular system. The new chapter in Jakob's life had started with a big 'Yes' to life. He was now five minutes old. But his heart had been beating for six and a half months. And his life had started nearly eight months ago.

The birth of the heart

At Jakob's procreation, his father's semen dissolved into his mother's egg. From this moment, every human being is genetically a unique individual who has never before existed and never will again. The inseminated egg cell divides itself into so-called stem cells, and after twenty-two days the little heart beats for the first time. This big bang of life needs a special ingredient, discovered only a few years ago. Researchers at the University of Montreal found that something magical happens with undifferentiated embryonic stem cells when they are sprinkled with the 'love hormone' oxytocin: they join, transmute into heart muscle cells and begin to beat in sync.[1] The love hormone is thus the essence of the first heartbeat. Isn't that fascinating? I will examine this hormone and its effect on the heart further in the sections below titled 'The all-purpose glue of love' and 'Heartbreakers'.

No one would dissect children's hearts to conduct such research; as is often the case in medicine, the results come from animal experiments. But the same magic happens for human beings: the love hormone that gets the heart to beat is created by its own stem cells, by the expression of certain genes – and it also comes from the mother.[2]

When was your first time? Maybe on an ice-cold February night? On a mild May evening? When did your heart first beat? No one took a photograph of this important moment; it happened secretly, quietly, hidden deep within your mother's womb. It was the beginning of your life. The first twitch of your first heart muscles transported a little blood that became the stream of your life. Seven weeks after procreation, the heart is already functional. Blood supply and organ development during the next eight months depend on the heart. It is an infinitely soft first *boom*. 'A sound that comes from silence.' That is what the heart is called in an ancient language, Sanskrit: *Anahata*.

When is a human being a human being? That is a big debate in stem cell research. As soon as the egg cell has been inseminated, the human begins to grow, a process for which its cells have to divide and proliferate. At this stage, a cell can still become anything: an eye or a toe or the heart or the tongue. Hence the name stem cell – as all other cells will stem from them. They carry within them our complete genetic information, and researchers hope that this fountain of youth will allow them to repair and even grow organs and body parts. But should that be allowed? To create human beings artificially and use stem cells from the growing embryos as a spare-parts store? Or would that be murder? Are they humans, yet, at that early stage? Philosophers, theologians and scientists debate who is best suited to answer these questions.[3] However, the brain alone cannot determine such questions. Ask your heart; it knows the answer.

I put my stethoscope on Jakob's heart, closed my eyes and concentrated on entering this subtle world of noises. The sound of Jakob's heart was clean. That's what doctors call it when they don't hear anything suspicious. No whizz, rumble or machine sound. This heart was beating fast and determinedly. *Clean heart sound,* I would later note in my emergency doctor's report.

Jakob's first heartbeat was not the beginning of his life, but it was certainly an important moment for him and even more so for his parents. When will his heart beat for the last time? That will mark the end of his earthly life. Or will they pronounce him dead one day and take out his heart while it is still beating? Will his heart thus continue to beat, even when he is dead (braindead)? This topic, too, is hotly debated at present. According to many experts, death is defined as brain death.[4] Life, then, does not necessarily span between the first heartbeat and the last, as you may have thought until now. But when exactly does it begin and cease? Maybe Eternal Life 2.0 exists, and we only change our shape?

Unimpressed by all these questions, the little heart grows quickly. Seven weeks after procreation its development is finished. From about the eighty-fourth day, sonography can monitor its wellbeing with ultrasound waves. The moments when I saw each of my children's beating hearts for the first time are forever etched in my memory. It was love at first sight. In this black-and-grey ultrasound shadow play, most parents can only distinguish the heartbeat anyway. Nature protects the child so well that it is not easily accessible even for our most modern medical methods. We have only a small number of parameters to assess its development. Heartbeat and heart prosperity are among the most important.

With the increasing development of heart and nervous system, from about the sixth month, the baby's communication with its environment is established. It starts to kick and push. Its heart rate increases when it does so, and slows down again when it rests. However, the foetal heart rate not only changes through its own behaviour, but also with the psychological and physiological state of its mother. The baby's heart adjusts to the mother's day–night rhythm. The baby's heart frequency and number of movements will decrease when the mother is not well and, for example, she does not have enough oxygen in her blood. If mum is stressed or afraid, the little heart will beat faster. Lack of oxygen, sleep hormones and stress hormones are passed on through the umbilical cord, and within a few minutes the baby will react to them too. And every drop of alcohol, even a single one, can cause damage, as the latest studies show.

Due to this synchronisation of the average heart rate, which can last up to twenty-four hours, one can say that the hearts of mother and child are intimately connected.[5]

However, the fact that a veritable conversation can take place via their hearts is a new scientific discovery.

The language of the heart

Mother and child are different people; each of them has their own inner beat and their own 'heart voice': the heart rate variability. As already described, at closer inspection the heart rate is not constant, but changes from millisecond to millisecond and from heartbeat to heartbeat. These constant changes are chaotic and can't be predicted, at least not by our apparatuses and mathematical models. Mother and child, like all people, tick differently – or so we thought until recently. But they can also synchronise themselves down to the finest time scale. Towards the end of the pregnancy, from the thirty-sixth week, there are phases during which the two independent hearts follow each other from millisecond to millisecond in perfect synchronicity. They beat as one. That is infinitely more subtle and exact than the approximate alignment of the average heart rate that was known so far. In one study, during the six five-minute periods that were examined, thirty episodes of synchronisation occurred, with an average length of fifteen seconds.[6] That is extremely surprising, as mother and child are completely independent. They don't have the speedy connection that is the nervous system, and neither is the child directly connected to the mother's circulation. Substances are exchanged only through the umbilical cord, but this path would be far too slow for such a fast synchronisation. I was downright electrified when I read that. How can hearts put themselves in sync when they are not directly connected? Physicists and health professionals of Boston's famous Harvard University, too, thought about this phenomenon.[7] Could it be based on the principle first described by Dutch naturalist Christiaan Huygens in 1656? If you mount two pendulum clocks next to each other on a beam, they will tick in sync after a while. Back then, Huygens assumed that the two pendulums synchronised themselves via tiny imperceptible motions in the beam.[8] However, mother and child are not pendulum clocks, but rather tick irregularly, erratically, and each have their own beat. So imagine two 'clocks' that tell their own time from second to second but nevertheless become attuned to each other.

In biology, a synchronisation of independently oscillating systems has occasionally been observed elsewhere. For example, the synchronous chirping of crickets, the synchronous discharges of nerve cells in our brain (as when they comprehend language) or the simultaneous glowing of fireflies.[9] The sinus node, in the area of the right atrium, is a tiny node in our heart, so small not even a heart surgeon can see or feel it. All excitation of the heart originates from it, and it consists of millions of cells. Every single cell is a small oscillator, and before a heart gets to do even one *ba-boom*, all of the cells have to discharge together and synchronously.[10] But still, which secret mechanism leads to the perfect synchronicity of two hearts, those of mother and child? Imagine this: we do not know!

Possibly the ears play a role in this. The child can at this stage in its development already hear motherly heart tones and the whooshing knocks of her pulse wave. This could well be the missing weak impulse that drives the child's heart to beat in sync with the mother's. If one views the development of a human being as a continuous process, divided into a time before birth and a time after, the synchronisation of the heartbeats of foetus and mother may indeed be seen as the baby's pre-natal communication with its environment.[11] We already send our first messages as unborn babies in the womb. The language we use is the language of our hearts – for me, this is communication between two humans in its most fundamental, earliest, purest form. It is not two brains exchanging information, but two hearts.

The synchronous glowing of fireflies, the chirping of crickets and the simultaneous firing of our brain cells – it is always about communication and connection, there are messages encoded here. It is assumed that a mother's special awareness of her child is founded in the synchronisation of their hearts. Scientists found confirmation of age-old knowledge in their experiments: mothers know when the unborn child is well, how it is faring and if it feels poorly right now. This special awareness may arise from the communication via the heartbeat.

Another answer is to assume the mother's heart is the pacesetter. Or perhaps the baby's heart. Or both become attuned to each other in a very precise and synchronous choreography. The mother cannot hear the baby's heart, but maybe her heart can notice it – with seismological refinement – via signal paths not yet known to us. In systems that are in resonance with their environment, small vibrations may intensify to a single large amplitude – which may be so powerful that it makes bridges collapse, or small waves become gigantic ones. Would many hearts be capable of something like this? . . . And how was Jakob's heart?

He had become still. So still I was suddenly no longer sure he was okay. I examined him carefully. Then he started to move and grabbed my little finger as if to say hello. And he screamed again. But less strongly than before – it now sounded like whimpering. Maybe something deep inside him knew that the separation from his mother was life-threatening. He needed protection, warmth and food rather than a cold stethoscope on his chest. But he was not alone. His mother was here, impatiently waiting for the end of my examination. When would she be able to take her son into her arms?

'Jakob, now you can go to your mum,' I said and handed the newborn to his beaming mother.

She blew the fringe from her forehead and placed Jakob on her tummy. His face reddened a little, and he breathed more calmly. There was no other sound. He was safe now, and he sensed that. Lying on the mother's tummy is called 'kangaroo care'. This skin-to-skin contact is best for both mother and newborn. Their body temperature, breathing and heartbeat once again align.

The all-purpose glue of love

The connection between mother and child is one of the closest relationships that exists in life. Science and psychology call this relationship 'bonding'. This term means something like 'adhesion'. Parents, especially mothers, are thus glued to their children. Scientists avoid the word 'love', perhaps because it is so immensely powerful and multifaceted in its meanings. Scientists prefer definitions, as with them one can compartmentalise the endlessly complex world into manageable little concepts. In the world of science, parents don't love their children – rather, they are glued to them.

The glue consists mostly of genes that are responsible for certain types of behaviour, and of a variety of hormones. The most exhilarating ingredient in this magic potion is the love hormone oxytocin. For nearly 100 years it was believed that oxytocin was produced only in the brain. But for a few years now, we have known that the heart, too, produces the love hormone, and does so in rather effective dosages.[12] This hormone not only makes our heart start to beat, it is also important for its further development into a big, strong heart. That is why a little foetal baby heart produces even more oxytocin than an adult's big one. The heart's oxytocin can unfold its effects independently of the influences of the brain. It even seems to be the case that the heart can stand in for the brain and produce love hormones when the latter's production decreases.[13] Whatever the thing we call love may be, it really touches me that (purely biologically speaking) it is not only the brain that is an organ of love, but also the heart. The love hormone, after all, influences our feelings, our approach to relationships and our choice of partners. It not only serves as the starter for the motor of life, but also connects us to other hearts. But flexibly, not adhesively. That is an important distinction. In our hearts, every exertion is followed by relaxation, and in nature every connection is followed by separation.

The nearer the due date, the more oxytocin the mother produces – which awakens her maternal heart.[14] Simultaneously, the baby's heart, too, distributes more and more oxytocin. And now something hugely interesting happens. This sweet hormone intoxication leads to the two of them being able to let go of each other – the high oxytocin level triggers the contractions! Isn't that astounding? Doesn't it teach us a lot about the wisdom of nature? Love also means giving someone their freedom when the time comes. Oxytocin therefore really deserves to be called a love hormone. It connects us and also gives freedom and space for one's own breaths. And as with the heart and with all oscillating biological systems, relaxation follows exertion.

Jakob and his mother were now literally imbued with oxytocin, a state that also relieves pain, soothes the memory of an exhausting birth and creates feel-good sensations in the brain.

Some researchers conclude that the mother's brainwaves can synchronise themselves with the heartbeat of the baby – without them touching each other, only through the mother giving her attention to the child. They suspect the electromagnetic fields of both organs influence each other.[15] Looking at Jakob, I now even had the impression that his heartbeat was casting a spell over the brainwaves of everyone who was present. I relaxed, and the paramedics relaxed as well. I did something I rarely do as an emergency doctor ... in fact I had never done it before. Nothing. I waited and honoured the moment. Jakob looked nice and rosy and apparently felt good. My eyes now sought out the pale face of the father, who had held his wife's hand the whole time, bravely and quietly. Having experienced the births of my own children, I was able to imagine what he had been through. Now he had recovered somewhat and wanted to help a little: to cut the umbilical cord to separate for good that which had already been separated. His hand was shaking so badly I had to guide it. What was the state of his own oxytocin distribution at that moment? We know today that fathers, too, produce oxytocin when they touch their prematurely born babies.[16] And little Jakob was now out in this world.

'Everything went well after all,' the father said to me. 'Why do we still have to go to the clinic?'

'To be safe,' I explained. 'At the clinic, Jakob will be thoroughly examined again.'

'Should I pack a bag for these two?'

'Yes, please,' I said, and I too packed something – Jakob in a thermal blanket. A little king in his golden coat.

Heart to heart

He was close to his mother's heart for the first car ride of his life. The skin-to-skin contact saw the oxytocin level of mother and child rise further and stay elevated.[17] We sped through the night. As we were nearly the only ones on the road on this mild soccer evening, I switched the sirens off. Only the blue light threw its flitting shadows against the long alleys of windswept trees. The rough coastal winds had made these trees grow crooked. 'Who rides so late through night and wind?' Goethe's nightmare poem crept into my brain. 'My father, my father, do you not hear, what earl king whispers into my ear?' What the hell were these lines doing in my head? Everything was fine! Jakob and his mother were wrapped in warm blankets. I had both hearts in my view on a monitor. Jakob's life before birth had become his life after birth. He had left the element of water in his mother's womb and had surfaced to the thin ocean of air that coats our planet. It was a different, exposed, alien world for him – but one thing had remained the same: he could hear his mother's heartbeat. That calmed him. During breastfeeding, too, it is not only milk that relaxes babies, it is also the familiar *ba-boom* which reaches their ears. The *ba-boom* is reminiscent of the time in the womb. Newborn babies have an excellent memory of their mother's voice and heart sounds. It is almost unimaginable today that until fairly recently people believed the brain of a newborn baby was completely inoperative, as it was unfinished – empty, so to speak. But Jakob had as many brain cells at birth as an adult: about 100 billion. However, his brain weighed only a quarter of an adult's, as the nerve fibres were not isolated and not yet connected.[18]

His brain was thus not fully developed and did not yet facilitate little Jakob's thinking and understanding. Medical professionals call this state 'devoid of cognitive function'.

Some children are born without cerebral hemispheres. Neuroscientist Björn Merker describes the life of such children in a hugely interesting scientific piece entitled 'Consciousness without a cerebral cortex: A challenge for neuroscience and medicine'. Such children have an illness called hydranencephaly and are awake and conscious. The skull's shape remains normal, and the absence of the cerebral hemispheres is often not noticed at first. They only have a brain stem, and after a few months it becomes obvious how badly disabled these children are. However, they can live for years or decades with good care. They communicate with their environment in a simple way. They are particularly able to express their feelings. They rejoice, they laugh, and when they don't like something they cry and become sad. They are excited, sometimes curious and they can have a favourite toy. In short, they are conscious without a cerebrum. Their whole existence and their heart must play a role here, as these children have feelings and they love. What do they love with if they do not have a cerebral cortex? Scientists increasingly believe that primary consciousness develops on the basis of nerve impulses that originate in the whole body. They have their origin in all our cells and organs and, of course, in the heart.[19]

How does Jakob recognise his mother's voice and heartbeat? And where is his understanding for the code of her heart located? Hardly in his immature brain alone, even though some scientists believe the brain's parts are ready to the degree that a simple form of consciousness could develop.[20] That is controversial – but the colleagues only consider the brain and overlook the fact that the heart is fully developed. Even more, it is fully and directly responsible for its life; it has its own small nervous system and has for weeks communicated with the mother's heart, long before the brain can hold as much as a single thought. Is the heart therefore a special organ of knowledge and consciousness? Based on our observations of children without a cerebrum and on the heart communication of newborns with immature brains, I would deem this possible. I would describe this consciousness as 'heart consciousness', or scientifically as cardio-cognitive experience. In this, I see more than an extension of the well-known theories of neurocognitive consciousness – of which heart consciousness is the origin!

The brain governs our bodies – that is modernity's idea of humans. However, a baby's first heartbeat is not controlled by the brain, because at that moment it is not yet developed. Rather, it is the other way around. A functioning heart that can transport blood and nutrients is the prerequisite for the development of the brain. When the heart is already developed, the development of the cortex is barely beginning, in the eighth week. An EEG can detect the first sporadic brainwaves in the brain stem from the twelfth week of pregnancy and in the hemispheres of the cortex from the twentieth. The heart has long been fully developed by then and is used to monitor foetal wellbeing (eighty-fourth day). With the growth of the nervous system, the heart is the first fully developed organ that sends its signals to the first brain cells. Subsequently, the brain really explodes: 40,000 synapses per second develop from the thirty-fourth week onwards.[21] The voice of the heart thus contributes decisively to the development of the brain. There are only a few locations in the world where one can create a foetal magnetoencephalogram, which measures the little brain's electromagnetic waves. In this way, researchers record the first tender brainwaves; these, too, change with the mother's heartbeat.[22]

I am guessing that science has a few more surprises in store for us in this field. An intelligent brain will not forget the voice of the heart and will remain connected with it. The heart's *ba-boom* is not a side noise occurring when the heart valve closes, nor an acoustical aid for a doctor's diagnosis, but our first cradle music, our first voice and our life's first language. The first thing we hear is the beat of life. One could say that the first thing babies learn – or, better, *do* – is listen to the heart. Why, oh why, do we so often lose this ability in the course of our lives? It seems to be suppressed with the continuing growth of our brain by what we view as the brain's 'higher' cognitive functions. Thinking, planning, acting, doing. They are wonderful abilities, but by themselves, separated from the heart, they lead us further and further away from ourselves, from our source and our origin. Away from who we really are.

Hearts know more than we think

With relief I saw the brightly lit clinic appear after the last turn. I was noting Jakob's current results in my emergency protocol when he opened his big, blue eyes for a second. He looked at me as if I were from another planet. Penetratingly. Without expression. And yet as though he were looking deep inside me. Then he closed his eyes again. It was the first time after his birth that his eyes encountered light rays. In the womb it is mostly dark. A newborn baby's vision is blurry, they see shades of grey. They can only recognise things when they are close. What is important to them is sensing with their skin, and listening to the hearts and voices of the people they are attached to. Even if it sounds a little trite, the things that are most important for a newborn baby are invisible to their eyes. Jakob was pure sensation, comparable to a state of deep meditation. Newborns, they say, have a pure heart. I am convinced that what is meant by this is their freedom from thoughts, judgements and memories.

The nature of consciousness is one of the biggest mysteries of the universe. Sometimes when one does not find any definite answers, it is advisable to go back to the origin. To where it all began, to our procreation, to our first heartbeats, to our birth and the first days of our lives. It was there that I had found a gently sparkling diamond whose argumentative hardness could withstand many discussions. Hearts can synchronise themselves very early in life, and we are able to literally be like a heart. And when one heart recognises the other, this immediately influences our consciousness. Even when one is born without a cerebrum. One can only truly know something – not believe and not think, but know – with one's whole heart.

In the clinic, Jakob was received by paediatric nurses and doctors. Before I handed over the little king in his golden coat, I listened to his heart one last time. *Ba-boom, ba-boom, ba-boom …*

THE HEART IN THE INCUBATOR

There are cases I will never forget as long as I live. Jakob's is one of them. But he was not my smallest patient, Maria was. Where I come from she would have been called a 'hatchling'. Born in the twenty-fifth week of pregnancy, she weighed only 580 grams. He heart made a machine noise. *ShSh-ShSh-ShSh-ShSh* – like the locomotives in old Sherlock Holmes films, or the steam engine belonging to my childhood friend Johann. He was a doctor's son and had this gadget I was really keen on. Unfortunately we were only allowed to play with it when his father was at home. 'It is dangerous and could explode,' we had been warned. I wished for nothing more than a steam engine of my own. But it cost more than 100 marks and was far too expensive. Not even my grandma could be persuaded, but at least she had an idea how to take my mind off it. 'Come with me, I want to show you something. We have hatchlings again.' She opened the big heavy latch to the hen house and let me slip past her long skirt into a mysterious world of golden-yellow flakes on two legs with orange beaks. The little bundles were huddling close to each other but also constantly on the move. Above them was the sun of the heat lamp so they wouldn't jitter. Temperature needs to be regulated very carefully for both newborn chickens and newborn humans. They rely on warmth, ideally their mother's. As a progressive 'hen mum', my grandma hatched the eggs in an incubator and warmed the hatchlings under the lamp.

A few decades later, little Maria lay before me in an incubator. Very soon we would operate on her. She seemed to me like a delicate plant in a greenhouse. My colleague Yücksel, who would be my assistant, used a different metaphor. What he saw reminded him of half a meatloaf sandwich – a description that outraged a medical student who worked in the intensive care unit, until she realised that Yücksel was teasing her. He was known to never miss a chance to flirt.

Yücksel wanted to get a cappuccino before the operation. We had about forty-five minutes. 'One for you, too?' he asked me.

'No thanks.' I stepped closer to the incubator and whispered: 'You poor hatchling.' Such children incite tender mercy, as my grandma would have said. Maria's skin was delicate like freshly curdled milk, her blood vessels gleaming blueish underneath. Tiny electrodes were attached above her heart, which would need to be placed elsewhere when we opened her chest. The child was a premature little human being, weighing just over half a kilogram. She was not yet able to breathe by herself; she got her oxygen through the tiny lens tube that protruded from her mouth and went down the windpipe. Next to the respirator stood several small blinking medication pumps that dribbled life-sustaining substances into tiny catheters. Above it all towered a monitor, bigger than the child, which told me two things: her blood pressure was alarmingly low, and her blood had too little oxygen. Maria still had the circulation of a child whose home is the womb. A foetus lying in amniotic fluid cannot breathe. Its blood, flowing from the heart, bypasses the lungs via a special blood vessel. This vessel, the *ductus arteriosus (ductus Botalli)*, was the problem. It had not closed after the birth – which happens automatically within hours or days with healthy newborns. For Maria, several attempts to do so with medication had failed. She became bluer and bluer, weaker and weaker, suffering from a wrong connection from the aorta back to the lung. Medical professionals call this a left-to-right shunt, and it causes a steam-engine noise. The ductus is a rather large vessel, and its failure to close meant Maria's blood did not reach her organs, as it should, but was circling between heart and lung. Too much blood in the lung impedes oxygen circulation in the alveoli. Pressure is too high in pulmonary circulation, while for the circulation in the rest of the body it is too low. Maria had not been in this world for long and was very, very sick and more delicate than a raw egg. The slightest tremor could cause her tender blood vessels to burst, as had already happened once; the little one had already suffered a cerebral haemorrhage. That was why Yücksel and I had come to the children's clinic with our mobile operation team, in order to carry out the required intervention.

Our paediatric nurse, Huyen, was preparing Maria for the operation.

'What is a hatch—', she asked, as she was always looking to learn new German words. She had been in Germany only four years, but her vocabulary was already impressive. I explained it to her while she lifted Maria from the incubator and placed her under a large heat lamp. We would also operate under the lamp.

'Hatchling,' repeated Huyen as if to commit it to memory, and it sounded nice the way she said it. Yücksel thought so too, having returned wrapped in a coffee cloud, but he didn't know what it meant. Huyen explained it to him. Then Yücksel told us about the challenges of Kurdish poultry-raising: birds of prey. He spread his arms, planted himself in front of Huyen and whispered, grimacing in a threatening way: 'They don't only take chickens away but little children too!'

Huyen pushed him aside. 'Nutcase!'

'And sometimes even beautiful women!' Yücksel laughed.

Huyen took a deep breath. Were these two about to quarrel again? I had observed that a number of times already. No, this time Huyen only complained to the big chairman in the sky: 'Lord, please give my colleague some brains.' And then we all concentrated on the operation during which the ductus was to be closed.

Operation in the doll's kitchen

As the cut will be between the ribs on the left, Maria is placed on her right side. Yücksel is concentrating hard now and reacts attentively to my every cue. We have carried out this operation together a few times, and he knows how I like to work and how he can best help me – and he kindly adjusts to me. When I operate with him, I have four arms and four eyes. Four hands can operate on a patient of thirty centimetres only if their owners aren't shy of bodily contact. We stand as close as possible. Otherwise it would not be possible to look into the tiny body I was about to open.

'Knife.' Everything now follows a carefully rehearsed choreography. Theatre nurse Klaus places the knife in my open hand with slight pressure while my eyes are locked on the marked incision line between the ribs. I can't see Klaus, as Yücksel is standing between us. It is quite an art to place a small knife in an open hand so the surgeon receives it safely and it does not fall to the ground. Klaus does it perfectly.

Yücksel lightly touches my forearm and then looks me in the eyes intensely. In this way he signals that he is completely ready now.

Surgeons can read eyes as if they were open books. They look into eyes every day for many hours, and I believe it is correct what abbess and naturalist Hildegard von Bingen observed 1000 years ago: the eyes are the windows to the soul. We are otherwise fully covered in headgear, surgical mask and sterile clothing. Surgeons wear magnifying glasses, and I am also equipped with a head light, like that of miners who work in the dark. But we look similar to them sometimes, and every now and then patients even say to us: my heart is sitting in my chest like a stone.

It is now very quiet in the theatre. Blood is streaming from the five-centimetre cut. I sever the fine respiratory muscles between the ribs and open the pleura, the membrane lining the inside of the chest wall. Tiny hooks in Yücksel's hands, so small they could be from a doll's kitchen, hold the wound open. I insert the rib spreader and slowly wind it open. The lung becomes visible; Yücksel gently pushes it to the side. It is in a dangerous position, as breathing now becomes even harder. But a window opens through which we can see the *ductus arteriosus*. There are numerous blood vessels and nerves running through the chest, and it is essential to identify them correctly. With the tweezers I indicate all the important ones: the aortic arch, the *aorta descendens* (descending aorta), the *arteria subclavia* (subclavian artery). Yücksel confirms this and points to the *nervus vagus* (vagus nerve) and the branch of it that is central to our operation, the *nervus laryngeus recurrens* (recurrent laryngeal nerve). Heart and brain exchange messages via the vagus nerve. The aforementioned branch does an extra lap around the *ductus arteriosus*, which is close to the heart, before ascending to the vocal cord, keeping it open. Without the *recurrens,* which means something like 'the returning one', we could only speak with trouble and produce only croaky sounds, as the glottis would half-close.

After Yücksel and I have identified the anatomical landscape before our eyes, we set about the task of closing the *ductus arteriosus*. The blood vessel made from embryonic tissue is only half a centimetre long. Its walls are not elastic and flexible like those of other arteries, but extremely prone to tearing. It is not made for a whole life but meant to close and become a strand of connective tissue that is still present in adults and called *ligamentum arteriosum*. The operation consists of extracting the *ductus arteriosus* from the surrounding tissue, entwining it with two threads and disabling it with two knots. Quite simple, really, if it weren't for its tininess and tenderness.

The heat lamp is shining onto my head, which feels as if it were in a toaster. I am one of those people who remain at a steady temperature even in a sauna. Yücksel always sweats, so he wears a headband. While operating on the extremely fragile *ductus arteriosus* I feel as if I were part of a bomb squad kneeling over an explosive device. Extracting the detonator from the centre isn't usually a big deal. But it needs only a tiny tremor, the smallest false move – and an explosion will occur. In my case I would then see red, as in a split second the operation area would be flooded with blood. Even a tiny person has a rather big aorta and big lung arteries. I made tiny cuts with fine scissors around the *ductus*, the path connecting these vessels. Incise a little, spread a little with blunt scissors, but not too much. Drops of blood continue to seep from the surrounding connective tissue. Yücksel has the doll's kitchen suction drain ready and at the same time touches my tweezers with an electronic cautery device which coagulates the injured blood vessel caught between its jaws. The anaesthetist asks us again and again to let go of the lung as the oxygen supply drops too much. Then the whole area of operation vanishes beneath the inflating lungs and has to be readjusted. Finally we have uncovered the *ductus*. I pass a clamp under it, then gently open the clamp just enough to allow Yücksel to place a thread into it. It is a fluent, synchronous movement of two people with four hands. We work and breathe in sync. It is highly probable that our heartbeats have synchronised.

Surgeons in the choir

When two people master a difficult task such as working together in surgery, playing in an orchestra or dancing together, a large amount of intellectual and physical choreography is required. Heart rate variability provides a parameter, for the first time, with which to determine how hearts beat in sync. I was quite surprised to find that the hearts of mother and child synchronise themselves and that the child's hearing possibly plays a role in this. But what about two grown men in sync when they operate on a heart? So far we have only one report on the interlinking of chaotic oscillators such as heart and breathing. It was published by the renowned Royal Society in London.[1] With a lot of lifeblood and physics, electronic engineers and doctors succeeded in proving that the synchronisation of hearts in well-rehearsed teams will rise exactly when really tricky situations are being mastered. This phase-synchronisation of hearts was also described for singers in a choir, which seems logical to me. When they sing a piece, they have to breathe nearly identically. Coordinated breathing is possibly a catalyst that synchronises other oscillating systems such as the hearts of the fellow singers. The slower the breathing, the better the mechanism of this so-called resonance frequency breathing will work.[2] Coherence between the hearts of adults occurs especially strongly during the singing of mantras and hymns – which may shed new light on the singing of soccer club songs before games. Could team members become one in this way to focus on the task ahead? Or are the individual players too nervous before the game? The origin of the lyrics, by the way, does not matter. Coherence occurs during 'Holy God, We Praise Thy Name', as well as during 'Om Mani Padme Hun' or collective yoga breathing.[3] Scientific studies show that during a collective activity we abandon our egocentric worldview in favour of a common perspective.[4] Many of our inner and outer senses play along in this concert of coherence. The lung's stretch receptors, the autonomic nervous system, receptors in the heart, our hearing, our hormone system, our muscles and movements and of course our brain attune themselves to each other and dance together. Collective activity and experience leads to common perspectives and goals.[5]

Happy end

'Should I dab the sweat away?' Huyen's question is intended for Yücksel, and he turns to her thankfully. Some more hearts seem to synchronise in this intensive care department. Yücksel and I continue to stitch. It is the knots that are critical. If they are too tight they can damage the vessel; if they are too loose the *ductus* may not be properly closed.

'Saturation and pressure rising,' reports the paediatric anaesthetist – so the blood's oxygen content is going up together with the blood pressure. Then he vanishes with his ear trumpet behind the the sterile drapes separating us from the anaesthetic equipment. After two minutes he pops up again: 'The machine noise can no longer be heard.'

Pfffffffft.

My tension releases itself in a concentrated exhalation. *Pfffffffft* is the sound made when maximum tension escapes from the lungs, like a locomotive. I have often heard it in the operating theatre. With some stitches and sutures, you only have one chance.

The rest of the operation is routine. Carefully we check that everything is 'tight', as we call it. As I am very meticulous in this regard, Yücksel jokes, not for the first time, that I'm counting red blood cells. He looks at Huyen for approval. She smiles. The strain and the worry for Maria have vanished from her face, have vanished from all our faces, or at least our eyes – which is all we have seen of each other for a while. Finally we put in a drain tube, made of flexible plastic, for fluids from the wound to flow through. Then we slowly retreat, stitching all levels of the wound carefully. Only a small scar will remain.

I hope that Maria will soon be on her mother's tummy and hear her heart. When that is not possible, recordings of the mother's heart sounds are sometimes used to calm babies in intensive care.[6] Close to the heart they feel at home.

I take off my green clothes and look forward to the coffee Yücksel, Huyen and I had planned to have together. After all, we haven't sufficiently discussed the topic of hatchlings. But to my surprise I see the two of them setting off without me. And judging from how closely they are walking beside each other, I don't think they will miss me. A few minutes to myself now would also do me good. I am still under considerable strain, even though the inner tension is gradually subsiding. And my heart supports me in this.

WHAT THE HEART CAN FEEL

The heart has a fine and sophisticated sensor system for the pressures in its ventricles and atria. It knows how much it is stretched and even has sensors for the chemical environment and the composition of the blood in its chambers.[1] These data are passed on to the brain, and it depends on them how much blood and information is ejected with the next heartbeat. It becomes too much sometimes even for a robust type like our heart. If there is too much pressure and it is all too 'full', the heart may let off steam. It spills hormones called ANP (atrial natriuretic peptide) and BNP (brain natriuretic peptide). We have to take a break, go to the loo and urinate, the amount of fluid in our body decreases, the blood vessels widen and the excess pressure in the heart abates.[2]

ANP reduces the amount of stress hormones circulating in the body and exercises a positive influence on our immune system and our behaviour.[3] The heart's ability to perceive physical signals allows it to always return to an equilibrium, to a balance of exertion and relaxation. Some scientists say the heart acts intelligently.[4]

Heart breakers

Had Yücksel and Huyen already finished their coffee break? I couldn't find them in the cafeteria. Only when I was looking for a seat with a cappuccino on my tray did I spot them. They had retreated into the furthest corner. To fight? No, not this time. They were enjoying an animated conversation. And it looked like much more than flirtation. I decided, grinning, not to disturb them but sat down not too far away. They did not even notice. I enjoyed my coffee and let my thoughts wander.

With couples in love, sometimes a look into each other's eyes is enough for their heart and breathing frequencies to align.[5] Yücksel and Huyen were now sitting so close to each other that my clinical glance immediately detected the cuddle hormone oxytocin. In the section 'The all-purpose glue of love', above, I discussed how the heart is able to produce oxytocin by itself. Thus it is not surprising, even though not widely known among heart specialists, that our heart is equipped with feelers for the love hormone. Those who believe that love originates in the brain alone may well ask why, if this is true, the heart also has such receptors? Our body is a highly complex, finely tuned masterpiece – every receptor and every pipe has its purpose, nothing is left to chance.

As you read earlier, the love hormone is the initiator of your first heartbeat. But it also has a life-long protective function for your heart. Researchers say it is cardio-protective: it constrains the distribution of inflammatory mediators and thus the development of atherosclerosis. It has been shown in experiments that the amount of dead heart tissue – indicating the size of a heart attack – was reduced by two-thirds if the hearts had been flooded with oxytocin, as it reduces the heart's oxygen requirement and widens its blood vessels. Blood pressure drops and the pulse becomes slower.[6] In stem cell therapy involving the heart, researchers let new heart cells grow when the old ones have been destroyed during a stroke. Before the new ones are implanted in the heart, they are bathed in oxytocin, because it allows heart cells to grow especially well.[7]

Several studies have found that people in stable partnerships live longer.[8] I am sure that has to do with love and its repercussions for the heart. And it might well be that Yücksel and Huyen were laying the foundation for a long happy life right at this moment.

Yücksel kept interrupting his animated chat and touched Huyen's upper arm or left hand as if by accident. As if he wanted to be sure she heard his words and opened her heart to him. What is vital for us as babies we also like as adults. A loving touch can synchronise romantic partners in an instant. Their skin resistance, breathing and heart rates attune. Even the time the pulse wave needs to travel from the heart to the tip of a finger adjusts.[9] Recently it has been proven that the skin possesses special nerve fibres, so called C-tactile afferents, which trigger the release of love hormones.[10] My two colleagues looked as if it all came from the heart, with the signal paths via the brain playing only an inferior role.

With each one of Yücksel's touches, Huyen blushed a little, and Yücksel's cheeks were already glowing. Soon the lights in the cafeteria might be switched off. Possibly the two of them would not even notice; they had not once looked around the room, had not seen me, for example. Well, I had no part in their romance. What I saw with increasing amazement and great joy was a most beautiful spectacle – what people have thought–felt–known for thousands of years: the heart is the organ for love. It beats in our throats and races with excitement and joy. It flutters in the chest and becomes big and wide. And sometimes it is given away, as was happening five tables down from me. We blush and our blood pressure rises, the hands shake and sweat. The electric skin resistance is lowered, the skin becomes more sensitive, we become literally more touchable.[11]

A love potion made from adrenaline, noradrenaline and the joy hormone dopamine is responsible for these unmistakable symptoms. Its ingredients are well-known: neurotransmitters in the brain and hormones from the adrenal gland. That they also come from the heart and are produced there by specialised cells – that is a fact so far completely neglected by love research.[12] Instead, love has been more and more vehemently situated in the brain in recent years. That did not happen by chance. It is in the brain that noradrenaline causes sleepless nights and lack of appetite. Dopamine has inspired poets to write love poetry and musicians to write love songs.[13] I believe that it is precisely the *duet* of heart and brain that achieves great works of art. And a few tables away I was able to observe lines from many well-known love songs in action.

I took a sip of coffee and thought of my little patient, Maria. In a minute I would go to the intensive care department and check on her. The biochemistry of love and the therapy to save gravely ill hearts in intensive care are basically identical. For the heart not only produces noradrenaline, adrenaline and dopamine, it also has receptors for them.[14] Heart surgeons know this, and after heart operations these substances are given intravenously for days and weeks to support the gravely ill heart. I doubt, however, if 'gravely ill' would be the correct diagnosis for the hearts of Huyen and Yücksel, even though they had begun in the meantime to appear a little feverish – and also disoriented, as actually our break had come to an end.

Hearts don't lie

In one study, the partners of thirty-two ethnically mixed couples in love were placed across from each other, and their breathing and heart rate variability were measured. Six per cent were Asian, 75 per cent light-skinned American, 2 per cent of African descent and 18 per cent Latin American. First, they were asked to look deeply into each other's eyes for three minutes. No touching, talking or grimacing. Their hearts followed a shared trajectory of fast to slow, then becoming faster again nearly at the same time. Independent of the participants' origin, Cupid's love beat seems to pulsate in a universal language that is understood and is the same everywhere.[15] In a second round, the participants were asked to try and imitate their partner's breathing and heart rate. Concerning the breathing this worked quite well; it is visible after all. But with the hearts, the subjects lost their beat; the effect was partly even the opposite of the intended one. For example, the male hearts beat faster when the female hearts were slowing down. In other words, the participants were not able to deliberately establish this sort of touchless beating in sync. At the risk of sounding corny, I say: hearts don't lie.

The ringing of the telephone tore me away from my thoughts. My mother was calling. My father had not been well for some time. My parents had been married for over fifty years. They were, so to speak, at the end of the path that Yücksel and Huyen may just have started together.

'Can you come home soon?' my mother asked me. 'Papa is not feeling well. The doctor says his heart is getting weaker and weaker.'

DANSE MACABRE

Some days later I was on the train home. Maybe this would be my last visit to see my father. A few years ago, he had had a new heart valve implanted by a surgeon who was a friend of mine. Dad and I had talked about his life, and about death, before and after this procedure. He had helped create me many years ago; perhaps soon I would hear his last breath. You might wonder why I did not operate on my father myself. It's because emotions are unpredictable. Even for surgeons. The fear and pain would have been too great and might have diminished my matter-of-fact way of functioning, which would have been so vital.

Dying is a spoke in the wheel of life. Humans die permanently and are reborn every second. Fifty billion cells perish in our body every day and are replaced by fresh ones. Statistically speaking, every single cell is replaced within a year, as well as

98 per cent of atoms and molecules.[1] However, we do not perceive this programmed cell death; we remain the same, after all. Or so it seems. But our soul or self changes throughout the course of our lives, due to perception and knowledge.

In various spiritual traditions, the death of the ego – one could also say its transformation – is viewed as a high level of knowledge. Not all people reach this level of consciousness, and mostly the ego dies last. This is recognisable, every now and then, from the gigantic tombs with which some attempt to defy mortality. In paintings from the Middle Ages, death is depicted as a creepy skeleton; it dances through the crowds and takes whomever it wants. The death dance of modern medicine is more varied: clinical death, natural death, unnatural death, heart death, brain death and cell death are its main performers. Not all are in the cast for good. And one variety gives empirical science a hard time indeed: the phenomenon of near-death.

When is death real?

According to an accepted doctrine, a patient whose heart has stopped, whose brain is not supplied with blood and who is not breathing does not have any consciousness. Not all patients abide by this.

In a fascinating study, the experiences of 344 patients with cardiac arrests were examined. If the electromagnetic waves of the heart ECG flatline, it usually doesn't take more than ten seconds for a person to lose consciousness – and in the brain, too, the EEG can no longer detect electromagnetic currents. As the term 'loss of consciousness' expresses, one would then expect that everything becomes dark for this person. However, for 18 per cent everything became light, and they embarked on a journey: they *experienced* how they left their bodies – combined with a very pleasant feeling and the knowledge that now they were dead. Some approached a light through a tunnel, saw colours and celestial landscapes, met dead relatives. Others saw their whole life like a film or found themselves at a kind of frontier. None of them passed the frontier irrevocably: all were successfully revived. Therefore they had not died but had had a near-death experience. On our side of this frontier, relatives, paramedics, nurses and doctors were fighting for the patient's life. Dutch cardiologist Pim van Lommel was interested in what patients had experienced at this frontier, and recounts this in renowned scientific journals and his book *Consciousness Beyond Life*.[2] Based on current estimations, 4 to 5 per cent of the Western population have been granted a glance into another world. Ultra-modern high-tech heart medicine seems to open a window for us onto otherworldly dimensions. It becomes ever more successful at bringing moribund, critically ill patients back to earthly life when they are at the threshold. At the threshold of what? Of the hereafter, of another world, of new dimensions? Some scientists believe the 'apparitions' are merely the final convulsions of the dying brain. But what happens if a brain specialist experiences something like this themselves? It happened to American professor of neurosurgery Eben Alexander, who had been in a deep coma with severe meningitis. In his book *Proof of Heaven: A Neurosurgeon's Journey into the Afterlife*, he describes his near-death experience, which lasted several days.[3] Sometimes this state occurs after an accident with severe loss of blood or a serious head injury. Even children who have nearly drowned speak of it, and occasionally a dying person may experience a visionary glance into the afterworld from their death bed.

For many people, dying is a phase of life, a transitional stage in which the spirit detaches itself from the body. Our bodies, bones and organs are made from the elements of the earth and return there. After the process of dying we are dead. But for many the spirit, the soul or pure consciousness lives on as a continuum without beginning or end. Scientists such as Eben Alexander and Pim van Lommel also arrive at this conclusion. Yet it is not at all new. It can be found in the age-old teachings of Christianity, Islam and many other spiritual traditions. There are different ideas as to which 'visa' the spirit will be given and where the journey onward will lead.[4] However, free entry is guaranteed for every human being, independent of their gender, race or origin and disregarding whether they are poor or rich, good or evil. But what and where are we once the heart is silent?

People's near-death experiences can shed some light on this. Their reports betray astonishing similarities across nations and cultures. In the aforementioned study, they perceived the 'celestial' journey as something very beautiful. They were interviewed again after two and eight years and compared with people who had not had near-death experiences. Although the experience had in most cases not lasted more than a few minutes, the survivors remembered it very well even years later. It had profoundly influenced their further lives. They were better able, afterwards, to show their emotions. Love, compassion and spirituality had become more important for them. People who had a near-death experience spoke of their increased belief in a life after death, and the majority had lost their fear of death. They had experienced that their conscious being did not cease when their brain and heart stood still.[5]

A matter of life and death

Some patients who are revived after a cardiac arrest have to be operated on immediately; this is extremely risky, as after an infarction sections of the heart muscle aren't supplied with blood and start to die. The heart has no reserves left and no power to supply itself and the other organs with sufficient blood. The patients suffer from oxygen shortage, their cells die en masse, so the whole person does too. This stage is called cardiogenic shock. The deadly downward spiral has to be stopped as fast as possible if there is to be any chance of survival.

In an induced coma and with heart medication at high doses, the patient is quickly wheeled into the operating theatre. The heart disease in question here is usually far advanced. Heart surgeons try to restrict themselves to the bare essentials in such cases. The longer the cardioplegia lasts during the operation and the longer the connection to the heart-lung machine is required, the graver the repercussions will be for a body already down for the count. Sometimes bypasses and a new heart valve have to be implanted. Or a new aorta straight away, as I described in the chapter 'Heart on the Table'.

Statistically speaking, the mortality rate for heart operations is very low. On the operating table, only extremely severe cases end in death. For example, an acute cardiac arrest involving a tearing of the heart wall, or sometimes an accident victim with grave injuries to the heart and lungs whose vessels have ruptured. I remember a young woman who was crushed in her car during a horrific accident. She reached the ER of the clinic alive, and my colleagues from emergency surgery attempted to stem the life-threatening bleeding via an operation. But she did not improve. What had not been clear at first glance was that the heart and the large blood vessels near the lungs were also injured. I was called into the theatre. Specialists from different fields were examining a heart which was only twitching weakly. The shock room resembled a battlefield. Long metal vessel clamps protruded from the open chest. The connections of heart and lung had been partially severed, and the patient's pupils were wide and fixed – a sign that her brain, too, was no longer responding. For this patient – and for quite a few others – there was nothing I could do.

When a person dies during a heart operation, a loud silence sets in, in a manner of speaking. This silence does not really seem quiet, even though we switch off the devices and the humming and beeping and the clatter of stainless steel ceases. Rather, the sounds from our fight for the patient's life seem to hang in the air still. The team will remain standing at the table for a while. According to their individual disposition, they mourn, feel exhausted, disappointed, empty or furious. We have given our best, but this patient's fate had something else in mind.

To see a heart flutter hurts me. But to see a heart die really gets to me – and is, at the same time, a spiritual experience of finality. Some hearts announce their death through fluttering, others beat to the last moment. The heart will become slower and slower, weaker and weaker. But it is still beating. And then it seems to me that the weak beating becomes more peaceful. Sometimes it gives the impression of giving up or agreeing to go. We who stand around the open heart in a circle, we follow and let it go. At this point we will have done everything humanly possible to save the patient; we will have fought and tried again and again to wean the heart from the heart-lung machine, three or four times. But without that support it took a turn for the worse, became too weak to support the body, too weak to look after itself. And so we disconnect the body from the heart-lung machine, knowing it will die. Such a death announces itself. The heart beats and pauses and beats again, and then the pause before the next beat becomes longer. Its beats are weaker, and the pause becomes longer still. And then it will beat one more time and never again.

If there is even a small chance for the heart to recover, we connect the patient to a permanent mini heart-lung machine; this is called extracorporeal membrane oxygenation (ECMO). It is an extremely invasive measure with many possible complications, as the patient is attached to a machine for days or weeks and bleeding, infections and strokes may occur. The use of ECMO is discussed by the whole medical team, as it has far-reaching consequences from both an ethical and an intensive care perspective, and many questions have to be answered. Does it make sense to keep prolonging the patient's life? Can we really improve their chances of survival? Such patients often receive massive transfusions, they are connected to dialysis devices, require artificial respiration and are kept alive with medication of the highest possible doses.

Above it all towers the question: how will the brain react? Has there been a resuscitation, and was it effective enough for brain damage not to be likely, or will the patient, should they survive, be severely disabled and become dependent on care? Would they have wanted that? If, after sober consideration, we think a patient has a chance, however small, we will take it. Even if we see the possibility that the patient may die after a few hours or days in intensive care. Heart surgeons are advocates of life; it is their task to save lives even when the situation appears hopeless. They are always the last resort. After us there is no other medical possibility. In competent care, patients may survive such extreme interventions. And then it will all have been worth it!

The decision to let a patient die is not made by the surgeon alone, but if at all possible together with experienced colleagues whom I ask to join the discussion. Perhaps they have another idea, a suggestion? If I operate at night and no one is in the clinic, I seek the advice of my assistants and anaesthetists; personally, I do not think that is a bad thing, since they have followed the course of events from the beginning. They have a connection with the patient's heart, while someone who is called in and informed about the case does not know its history, or else only its medical aspects – and that is not the whole truth. An open heart involves so much more.

It may sound strange, but I feel closer to the dying heart than to the patient. I have been in touch with the heart. I have got to know it. I recall its face much more clearly than the face of the person, with whom I may not even have spoken before the operation, or only briefly, as it was an emergency. I do not know them. But I have come close to their heart, have advanced into its deepest cavities. And it is their heart which I say goodbye to – a defeated farewell, because I have lost the fight against death. Usually I suture the chest myself in such a case, rather than leaving it to my assistant. It is a final service to the patient. There is a peculiar atmosphere in the theatre. It may be quiet, or someone has to tell a joke, or everyone becomes noticeably busy. We have nothing left with which to oppose death, we have to bear it, each in our own way.

The dead patient's age, of course, is a factor as well. A young person's death, or a child's, is hard to cope with. I remember a family man in his mid-thirties who was in limbo between life and death for weeks in the intensive care department. His plight touched us all, especially because his three small children came to visit with their mother on Sundays. It was a little as if they were reminding the team why it was worth fighting for this patient. On a mild August night, we knew that he would not see the next day, and called in his wife. On the monitor she could see her husband's heart becoming slower. And even though she was prepared for this, she was not able to just let him go. She beat her fists on his chest and screamed: 'You must not die! You must not leave me behind!'

That was hard to bear. I looked through the window, which was slightly open, hoping the other patients would not hear the woman's screams. On the windowsill was a bag of dialysis fluid, weighing about three kilograms. The dying man's heart became slower and slower. 'You must not go! You must not! Stay with me!'

After the last heart tone there was silence, and then a gust of wind threw the window open, knocking the bag into the room, where it exploded on the floor with a bang and caused a flood. I exchanged glances with a nurse. It seemed like we were thinking the same thing.

Beating heart cadavers

Nothing lasts forever, and a heartbeat only lasts a moment. Many moments add up to form your life. Most people believe they are dead when their heart stops beating. Today we know this is not quite true. Maybe they are only clinically dead. That is the case when a cardiac arrest is not final and resuscitation is carried out successfully. If that doesn't happen, all organs begin to die within five to ten minutes.

The brain can also expire of its own accord – in the course of severe illnesses or accidents: for example, through bleeding or meningitis. The heart is still beating, and the other organs are alive. Such a patient can no longer breathe by themselves and has to be given artificial respiration.[6] Doctors call this brain death. It is the latest performer in modern medicine's dance of death. It was first defined in 1968 as the final, irreversible end of all the brain's activities. Doctors may then testify to the death of the whole person and remove the organs of the beating heart cadaver, if he or she has previously agreed to this. However, it is also possible for a brain-dead person to continue to live for twenty years, connected to an artificial respiration device.[7]

The case of thirteen-year-old Jahi McMath caused quite a stir some years ago. There were complications after a tonsil operation, her heart stopped and she had to be resuscitated for two and a half hours. Her heart stabilised, but her brain had lost any detectable function. She was pronounced brain dead, and it was recommended to her family that they allow her organs to be removed. Her skin was warm and rosy, her face relaxed. Heart, lungs, kidneys, liver, pancreas and intestines were to be taken. After that, her artificial respiration would be stopped. But her family resisted. As long as her heart was beating, they saw Jahi as alive. There followed years of a lawsuit about the question of whether she was dead or merely regarded as dead. Four years after the operation, Jahi died from internal bleeding.[8]

Brain death is a definition, and when this definition is applied to a person, they are issued a death certificate. They are pronounced dead, even when (with the exception of the brain) all organs are alive and their heart is beating. In such cases, organs may be removed for transplantation. Behind this is the categorical will of visionary doctors to save human lives. Transplant surgeons take organs and tissue of the brain-dead person and implant them into other people. That is truly magnificent and has my full support. What is problematic, though, from my perspective, is the claim that the organ donors with a beating heart are dead. Not long ago a woman in Germany gave birth to a healthy child, after she had been pronounced brain dead two days earlier due to meningitis.[9] Her heart was still beating, her brain no longer working. In my opinion, the doctors acted correctly in letting the child be born and the mother die afterwards. But modern medicine poses new questions: when is dead really dead? Can the dead give birth? And how do you think about organ donation yourself – will you sign up as an organ donor or not? Would you bury a brain-dead relative who is given artificial respiration, while their heart is still beating? Or burn their 'corpse'? This last question is brutal and macabre, but demonstrates how much confusion the diagnosis of brain death can cause.[10] No one wants to be pronounced dead while they are still alive.

To relocate death from the heart to the brain is a scientist's heartless brainchild. And I believe that people all over the world sense this. Shouldn't the decision be whether I would gladly donate my organs when my brain has stopped working and won't ever work again? I will willingly give my organs, my cornea and other parts to someone who needs them and who will gain more quality of life with my help. However, I am not doing so as a dead person, but as a living one. It is my last good deed in life, not my first good deed in death. My time has come. This would be a clear, life-affirming decision. For this reason, I have an organ donor card. I decided this in full consciousness, and after my organs are removed I want to die 'conscious' of my decision. But what consciousness will that be if my heart is beating but my brain is switched off? According to the German Organ Transplantation Foundation, general anaesthetic is not necessary, as it would work predominantly in the brain – which is no longer working in this case, so no pain can be felt.[11] Moreover, the person has been pronounced dead before the transplantation, even though their heart was still beating. For this to remain the case, the whole spectrum of the therapies of intensive care medicine is exhausted to keep the corpse alive – including the neurotransmitters dopamine and noradrenaline, which also affect the heart. I don't know about you, but when I read something like this I deem it downright bizarre. If indeed I ever become an organ donor, I want general anaesthetic during the extraction.

As we have seen, there are children who are born without a cerebrum and who survive and are conscious, in a simple way. In the past, such children were used as organ donors – in the same way that premature babies were operated on without sufficient anaesthetic, on the assumption that they did not have any consciousness and they were not real human beings.[12] That is beastly cruelty and shows what people are capable of if they are led by reason alone, relying only on facts that can be objectivised and measured. As a surgeon, I consider helping people be free of pain one of the greatest achievements of medicine.

Who benefits from equating brain death with the death of the whole person? Some priests, doctors and politicians seem to think they can convince people to donate their organs if they make the well-defined brain death plausible to them. I believe the opposite is the case. It would be more appropriate to call brain death 'cerebral arrest' or 'brain failure'. In this case one should be able to give away one's organs and then die. In any case, different countries have different ideas about when exactly brain death occurs and what exactly has to be observed in order to confirm it.[13] This is comparable to the question of when life starts. If death equals brain death, does life start with the 'brain birth' – and when would that be? Only when the brain has fully developed, or when it is just beginning to grow? With the first brain cells from the twenty-fifth day after procreation, or shortly before birth in the thirty-sixth week of pregnancy?

Modern medicine does not know the whole truth about the beginning of life, nor about its end. Different views are debated, controversially, by science, religion and philosophy. However, if the transition and process of dying are finished and death has occurred irreversibly, doctors are able to determine it exactly and easily. Livor mortis develops, as well as rigor mortis and, in an advanced stage, decay. Those are the sure signs of death that every doctor today has to be guided by. That has not always been the case, and many people in former times were plagued by the fear of being buried alive by mistake, only appearing to be dead. That may have happened with a deep unconsciousness or a poisoning, when the pulse would have been so weak it couldn't be felt. The mistrust in medicine's competence regarding such existential questions has always been great. Time and time again, people who had been thought dead have awoken from their sleep, a blackout, a coma. The idea of being buried alive continues to be unbearable to many people today. In his book *Mein Leben mit den Toten* (*My Life with the Dead*), Alfred Riepertinger tells of a shocking method to overcome that fear, which even today is still carried out in Vienna, a city with morbid charm: the heart stab.[14] The dead heart is stabbed in order to give people who in life are afraid of apparent death the ultimate certainty they have asked for in their wills. There is a special heart-stab knife for this action – which is not deemed murder. The knife is about twenty centimetres long and sharp on both sides, as it has a double edge. Based on my experience with various knife wounds to the heart and those which missed the mark, I would claim that even the 'heart stab' is not 100 per cent reliable. Anyway, even Arthur Schnitzler (1862–1931), who was not only an eminent Austrian author of plays and stories but also held a medical degree, made use of this service. On the other hand, authors have vivid imaginations, and close to his death Schnitzler may have been more poet than doctor.

THE HEART IN THE EYES

When I arrived at my parents' house, I was extremely relieved to find my father conscious; he even had the strength to welcome me warmly. My mother and my siblings, who lived close by, had looked after him lovingly. I wanted to support them and simply be there with them. For the moments when I wasn't needed, I had brought along some work, including a scientific article from the journal *Nature Neuroscience* which had fascinated me when I read it the first time. Signals from the heart had been detected in the brain during a complex experiment.[1] Of course I had a keen interest in this and hoped to get an understanding of the experimental set-up; to do so, I would have to familiarise myself further with a field usually foreign to me, namely neuroscience.

After a few days with not a lot of peace and lots of family commitments, I had still not unpacked the article. My father was a little better, and it even looked as if he would be able to celebrate his eightieth birthday in a few weeks' time. I decided to take advantage of being close to the Brenner Pass and take a trip to Italy in the second week of my holidays. I visited Genoa, Portofino and of course Florence, where I had a long wait as one of many thousands of visitors to be admitted to the Academia. And there it stood, the best-known sculpture in art history – Michelangelo's *David.* It was much taller than I had imagined, even though I knew that David measured slightly over five metres. Of course it impressed me, but I found Michelangelo's many half-finished sculptures even more interesting. From them one could learn how he had worked and how mere stone had been turned into shape. They reminded me of the wonderful paper cut-outs by British artist Rob Ryan, who carved the following words out of paper to form the title (and cover) of his book: 'I thought about it in my head and I felt it in my heart but I made it with my hands.'[2] It seemed to me that in my search for the whole heart I would retrace this sentence back to front.

I would have liked to stay longer in the Academia, but the rush of visitors was so relentless that I was pushed past the masterworks as if on a conveyor belt, and suddenly I stood at the exit. There were Davids galore on offer in countless souvenir shops. Without so much as glancing at them, I elbowed my way through the crowd, stopped, and turned back to see a replica of David's head. No, I had not been mistaken. He had heart-shaped pupils! And this was despite the sculpture being a tremendously naturalistic depiction, with perfect shape and perfect proportions! Why had Michelangelo's genius deviated from this approach towards the centre of seeing, the pupils?

We see better with the heart

That evening, I became lost in the article from *Nature Neuroscience*, and it seemed to me as if a further piece, or stone, in my heart's puzzle was falling into place – not the one that David had hurled between Goliath's eyes, even though it had something to do with my sudden understanding. Little David, facing the giant, needed a lion's heart and could not afford to tremble when faced with such powerful superiority. Courage and love make one fearless, and the heart incorporates both qualities with its synthesis of adrenaline and oxytocin. But that alone is not enough. The young shepherd also needed a sharp eye in order to hit the giant – who was many times his size – with the utmost precision. His throw had to hit home, he had only one chance. And he succeeded, also because one's vision becomes sharper with a brain that listens to the heart – this was evident from the article that I had carried with me for so long. Now was the perfect moment to grasp it, with brain and heart.

The brain reacts to signals that come from the heart, and these signals influence our perception and our decisions.

With every heartbeat, the heart's sensors pass on information to the brain and the neocortex.[3] When the brain responds, so-called heartbeat-evoked potentials (HEPs) become detectable. These are measurable changes of certain brainwaves in response to the messages from the heart. To measure them is very involved and requires extremely sensitive technology.

In an experiment, subjects were shown a weak visual stimulus – grey-and-white patterns with such little contrast that they are right at the threshold of optical perceptibility and thus often are not consciously perceived. When such a stimulus was presented to subjects in a phase when the brain was reacting to the heartbeat, the probability that they could consciously see the image was significantly higher. Before the stimulus was presented, neither the multifaceted physiological body data nor the brain's general excitability could predict the correct visual recognition. When they saw the stimulus, the subjects had to press a button. The study's authors concluded that the heart's signals to the brain immediately influence our decisions and our behaviour.

Additionally, the heart's signals transport information to the brain that influences our physiological ability to see. If a stimulus is recognised, this process in the brain in turn influences the heart: after the decision it becomes slower and seems to relax.

This experiment betrays an enchanting, simple elegance. The stimulus was neutral, a pattern in shades of grey and white with little contrast. Emotions or certain expectations did not recognisably influence the brain's reaction to the heartbeat. The brain's responses were spontaneous, sometimes occurring and sometimes not. We do not know when or why the brain reacts to a heartbeat and when it doesn't. We can only speculate. This is also true for the message that Michelangelo carved into stone. Maybe it was an experiment, and the pupils in David's eyes, five metres up, could be seen only by those who were connected with their heart? That would be entirely possible, as current studies have come to the conclusion that we perceive ourselves more intensely and feel more compassion for others in situations where our brain clearly reacts to our heartbeat.[4] The amplitude, or strength, of heartbeat-evoked potentials is reduced when certain heart diseases and types of depression occur. That could mean that we are more susceptible to negative emotions such as feeling depressed, but also to heart disease, when our brain ignores the signals that are the voice of the heart.[5]

Such was probably the fate of neuroscience for a long time, as the brainwaves that have their origin in the heart are very fine and subtle. For decades it was believed that they were interference, as with a badly tuned radio. Today we know that it was not that the radio had been badly tuned; rather, our listening devices and measuring instruments were not sensitive enough. Extremely sensitive tomographs now show that the incomprehensible noise is in fact the sound of intricately structured body signals to which our brain responds and which influence our perception and decisions. In a scientific guest commentary on the study mentioned above, experts arrive at this insight: 'Given how seriously the brain appears to be taking the heartbeat, perhaps we as experimenters need to do so as well.'[6] Michelangelo took this to heart 500 years ago.

An interference entered my consciousness. It was my mobile. When I saw the number, I had a premonition. Thoughts about my father had been with me on this trip; sometimes it had seemed to me as if he was standing beside me.

'I will speak at the funeral,' I promised my mother. When I ended the call I knew that I would not give an ordinary eulogy. I wanted to speak with the voice of the heart, with the look of David in my eyes. My father had been a bricklayer. What would he have said about this sculpture? He had never been interested in 'things like that', but had not badmouthed them either and never put obstacles in my path, but rather had supported my development through the stones he laid and the houses he built. I had only grasped his savoir vivre late in life. And I demonstrated this three days later when I gave the whole eulogy in the wild Swabian dialect of our native area. 'Michelangelo would have depicted him with a heart in the palms of his bricklayer's hands. To build houses for other people was his way of expressing love.' My father did not respond that time; maybe he was already busy again in the afterlife, as his philosophy of life was 'Work hard, work hard, build a little house'.

A warrior's heart

Once I was back at my Baltic Sea home, I pulled a ten-kilogram tome on Michelangelo from the shelf. A friend had given it to me after my habilitation. The art world was ignoring this peculiarity of the pupils; I only found a few words about the hearts in David's eyes.

Instead, I discovered in the heavy tome a discussion among scholars about the question of whether Michelangelo had depicted his David before or after the fight.[7] I think both aspects are visible, and that is the artist's genius. He formed the white marble in a balanced state of exertion and relaxation, getting close to the nature of the heart. The sculpture depicts a physically present, focused David. He was only able to win this uneven fight in the state of a warrior who had gathered himself beforehand. In the moment of the fight he was connected with his heart, and his brain was open to the heart's signals – which gave him the strength and sharpness of vision to hurl the stone with all his might between the giant's eyes. This is where the 'third eye' resides – an energetic centre that represents visionary clarity and intuition, according to many spiritual traditions. In a certain way you kill a person if you extinguish these qualities. A person without knowledge, perception and intuition is dead inside. But perhaps David's fight was over, and the giant was at his feet. David's gaze has nothing presumptuous about it, and he does not boast; it is the gaze of someone who has done what he had to do in a straightforward manner. In the gaze of the man with the warrior heart there is compassion for the defeated and also the victor's gratitude towards those who helped him.

The collective heart

David was an underdog. But he had something that made him invincible. Not only his warrior heart, but also the hearts of those on his side who believed in him. He could be sure of the support of his people, the tribe of Israel. Hearts can synchronise themselves across vast distances and in dangerous situations. This was demonstrated a few years ago by people who literally walk through fire.

Every year on June 23rd, two tons of oak wood are set alight in an amphitheatre in San Pedro Manrique, a small village of 600 people. Three thousand visitors from near and far eagerly await the fire ritual, which begins at midnight. When the bed of glowing coal is ready, the procession of firewalkers from the market square to the amphitheatre begins. They are all inhabitants of the village, between nineteen and forty-six years of age. The crowd and their friends and relatives accompany them and cheer them on. Then they dance around the embers for a few minutes and walk barefoot across the seven metres of fire carpet, which is 677°C hot. Most of them also carry a person who is dear to them on their backs. In a spectacular study, the heart rates of twelve fire walkers (eleven men, one woman), one of their loved ones in the stands and other spectators (who had no personal connection to the fire walkers) were measured. Non-linear mathematical analyses uncovered that during the ritual a baffling alignment and synchronisation occurred between the dramatic-looking heart rate curves of the fire walkers and their loved ones. These curves are not at all linear, but like a rollercoaster of emotions. Every pair has their own highs and lows. In contrast, the heart rate of the spectators with no connection to the walkers remained indifferent.[8]

Gradually – and still largely unknown publicly – scientists are beginning to understand that hearts are able in multiple ways to connect synchronously with each other. But how exactly do they do that in an amphitheatre with several thousand spectators, and across quite a distance? After all, the heartbeat is not audible, the participants do not touch and they are involved in different activities. The authors of the study conclude that it would have to be a certain type of information that is exchanged between hearts, one we cannot measure.

In the frenetic collective excitement of a ritual, in the formation of a sense of togetherness, a palpable, physiologically measurable connection of hearts can occur; this affects the hearts of those actively involved as well as of those looking on. The supportive power of collective rituals is a phenomenon in all known human cultures. It fosters community coherence and encourages its members to stand up for one another.[9] We cannot deliberately influence our heartbeat, but – however it may work – hearts seem to sense when they are needed: they connect and go into resonance, similar to what certain nerve cells in the brain (the so-called mirror neurons) do in such situations. Compassion and support, then, are not only matters of the head but also concerns of the heart. The many scientific studies on the synchronisation of hearts allow us to conclude that the amplitude of many hearts may add up and become a collective heart; then we are like *one* heart.

LOVING

After my visit to Michelangelo and my father's death, I initially had trouble returning to the operating theatre. I would have liked to spend more time with what was dear to my heart, more time than the clinic's everyday routine would allow me. On the other hand, I was looking forward to the familiar environment of my daily work. So one morning I entered the nurses' room, which is always buzzing with intense discussions. I did not hear what my colleagues were talking about, but when I entered they fell quiet.

'What's happening?' I asked.

As if by command, everyone was suddenly busy. Irmgard was rummaging through papers, Klaus was preparing medication, Sarah was typing into her mobile, and two others made their way to the door. Only Bettina was grinning like a Cheshire cat. She had been the heart of the intensive care department for decades. With a big smile she told me the latest from the grapevine: 'Yücksel and Huyen are a couple!' she burst out.

'Oh, I've known that for a while.'

'You have?'

'Yes, of course.'

'Have they told you? How do you know? Klaus saw them smooching yesterday. You didn't hear that from me. Or did you catch them as well?'

'No,' I smiled. 'I diagnosed it from their hearts.'

Since time immemorial, people have entered into romantic relationships. Worldwide, 147 out of 166 societies have a word for 'love', and scientists assume the remaining nineteen possibly have just not been asked the right question yet.[1] While Adam and Eve in paradise displayed a rather earthly disposition, tried forbidden fruit and produced offspring, Mary and Joseph were more inclined towards 'immaculate', divine love. There might be a difference theoretically, but the practical result was the same. Reproduction is the true reason for the evolution of love, scientists say. During orgasm, we discharge a very special essence of the heart, the love hormone oxytocin. It is one of the oldest substances of life that we know of today; it was active on Earth over 600 million years ago.[2] Love is a universal, connective force. Some say it is very old and everlasting, the origin, the beginning of the beginning.

That may be the reason why lust and love are by no means absent in the later stages of life. Islamic scholar Hamza Yusuf once asked his 84-year-old father if physical lust ceases in old age. His father answered humorously: 'Yes, about a half-hour after the cadaver is cooled.'[3]

I believe that too, and current scientific inquiries confirm it. Love, physical closeness and sex are not only beautiful but also healthy.

A large study has shown that the frequency of sexual intercourse is an important factor for the mental and psychological health of both men and women.[4] Involved here is an old acquaintance from earlier in this book, which coming from the brain also visits the heart and travels further to the genitalia. It is the vagus nerve, the stray and vagabond. Even though it is a brain nerve, it wraps around the vagina and the cervix with its sensitive fibres and transports their messages directly back to heart and brain, without a detour through the spinal cord. That is why women with spinal cord injuries may experience orgasms.[5] For men, too, the vagus nerve regulates arousal. It makes arteries wide and allows blood to flow into the penis so that the spongy tissue fills and the penis becomes erect.[6] High oxytocin levels influence the data flow between heart and brain and make the vagus nerve flexible.[7]

When it receives sensual signals, the heart begins to dance, the heart rate variability rises; this has positive effects on our health.

The heart rate variability of 143 men and women was measured, and they were asked how often they had had sex in the last month. Women and men with a high heart rate variability had more sex and more frequent orgasms. Vice versa, orgasms increase the heart rate variability. Coitus increases the activity of the parasympathetic system – that is to say, of the vagus nerve – and the more relief they achieve, the more likely further meetings of penis and vagina become.[8] (Unfortunately, at the time of preparing this manuscript, I am not aware of any studies concerning homosexual couples.)

The voice of the heart, which finds its expression in heart rate variability, is a voice from inside the body – from the part of the body that has no other voice. It expresses what the tongue can't say, and it is the result of the complicated cooperation of various different organ systems. Strictly speaking it is therefore not one voice, but a multitude, a polyphonic canon of all cells. We humans have so far not managed to decode these voices completely. Expecting to decipher the heart's messages from an analysis of its frequency is as unrealistic as filtering thoughts from an EEG. As a scientist, I am astonished by the many voices that come through when one measures heart rate variability – a canon whose tones have such deep effects on our sexuality. These sounds cannot be played 'on demand'; we can only feel them. They create themselves, they stem from our very own biological existence. They are made anew in each moment.

The measurement of love and the rediscovery of the heart's involvement are very interesting scientifically, but they measure something that goes beyond what apparatuses can detect: being alive. Sexuality, love and the voice of the heart are highlights of life, but they are also changeable and cannot be categorised. If we lose them permanently, that can lead to depression, anxiety, eating disorders, overconsumption of alcohol and drugs, and personality disorders. Too much stress can be enough to hinder the experience of an orgasm,[9] as happened in Rüdiger's case.

The mechanics of love

Rüdiger was a good friend of mine, and a successful manager. Once every year we went hiking in the mountains or spent a weekend sailing on the Baltic Sea. During one of these sailing trips we were anchored in a bay and he started to have chest pain. The previous evening, he had told me that his wife had a lover. They had been married for nearly twenty years and had three school-age children. Rüdiger was terrified of losing his wife and made an effort to fill their love life with lust and imagination. But his wife desired him less and less, and the more he tried, the limper his penis became.

I was swinging in my hammock on deck and watching the night sky, where a spectacular storm was brewing.

Suddenly Rüdiger stood before me, both hands on his chest.

'Reinhard, I am not well. There is a strange pulling in the heart area.'

After taking a thorough medical history and conducting an examination with simple on-board equipment, a diffuse clinical picture emerged. I could not exclude coronary heart disease, meaning a narrowing of the coronary arteries. From my ship's pharmacy I administered a dose of Nitrolingual Pumpspray under his tongue. Nitroglycerin widens the blood vessels around the heart. While Rüdiger, still very pale, lay down in his bunk, I considered how dangerous this situation was. If I had learned one thing in my twenty years as a heart surgeon, it was this: one must never underestimate a heart. But chest pain can have many causes. I studied the sea charts, the wind and the weather and thought about how we might be able to reach a port. Unfortunately it was not looking good; we had crept deep into the Swedish archipelago, into a bay that was completely sheltered, as strong winds from changing directions had been predicted for the next days. All in all, the situation did not appear definite enough to raise an alarm. Following an inner voice, I decided to wait. Two hours later, Rüdiger crept from his bunk. 'I'm feeling better again,' he grinned.

'That's good,' I replied, relieved.

'This stuff is the shit,' he added, and pointed between his legs with unmistakeable pride. 'It still feels warm here.

I haven't had an erection like this in a long time. What was in that?'

With a smirk, I explained it to him.

'Isn't nitroglycerin an explosive?' he asked. He was himself again, no question, but I was not sharing his euphoria quite yet and observed him with a doctor's eyes. 'What about your heart?'

'Nothing. No pain. All gone.'

Heart and penis are an inseparable couple. Their arteries are of similar size, have a diameter of about two to three millimetres, and react to the same substances. The penis, too, fills with blood and discharges it again after ejaculation. In this it follows its very own rhythm of the tides of love, and yet it is inseparably connected with the heart.

Viagra, knight in shining armour for weakened male desire, was originally developed as heart medication to widen the blood vessels around the heart. However, a 'side effect' was far stronger. Viagra also widened the blood vessels in the penis and created staunch erections. Thus, its fame as a potency-enhancing drug came about, and it rescues many men worldwide from their distress.

The better the heart rate variability and the more relaxed the man, the more durable is his erection. This is due to the vagus nerve being active and opening the gates; the hydraulics of the erection are set in motion: maximal influx of blood with simultaneous minimal efflux. When men are stressed and believe they have to perform in bed, the heartbeat becomes rigid, the vagus nerve loses its desire and man's best friend droops.

The penis's biomechanical qualities also make it a desired object of investigation for engineers. Like the demands on the wings of a plane, those on the penis are immense: it must be flexible and elastic, but must not bend or break.[10]

When the dams break and everything flows, men as well as women reach a fulfilling climax. Maximum exertion and at the same time the biggest possible relief are united in a heavenly sensation. Hearts race and thoughts stand still. For a moment, 'Radio Nonstop Thinking' is switched off.[11] The ego is silent, the self ceases to exist in this union, we are pure self again – heart consciousness. Some also describe it as a state of deep meditation, as a transcendental experience. Buddhist traditions compare the climb towards the orgiastic experience to the eight stages of dissolution one goes through when one dies. The disintegration of the I, the release of all control, is sometimes also called a small death.

However, it is statistically highly unlikely that we really die when we have an orgasm. In one study, 536 patients who had suffered a heart attack were examined, and only three of them had had sex within the hour before the heart attack.[12]

A group of pathologists, too, examined 'love death' and concluded that physical love rarely ends in fatality. In 38,000 post-mortems, the pathologists found hints of a natural death during amorous play in only ninety-nine cases, or 0.26 per cent,[13] even though the heart has to make quite an effort when we love to a climax. The pressure necessary for the man's erection is four to ten times higher than normal blood pressure. Women, too, need an increased supply of blood from the heart when they engage in sexual activity.

The heart not only gives life in the form of blood, but also takes care of the physical pressure of desire, which increases with the degree of arousal. It ends in the uncontrollable, ecstatic, rhythmical contraction of our muscles in the genital area which we call orgasm. The heart also performs an unwitting rhythmical contraction. Every single heartbeat is an orgasm of life! And maybe after twenty-two days there will be a sound coming from the silence. *Ba-boom, ba-boom, ba-boom …*

Sick of heart and listless

If the heart is permanently sick, potency may drop. The risk factors for coronary heart disease and erectile dysfunction are the same: high blood pressure, stress, smoking, diabetes, lack of exercise and obesity. Heart and penis act in concert, and with the progressing illness of the blood vessels, the blood supply for both decreases. Sometimes, though, it is the other way around, and the missing erection can be a harbinger of a stroke or heart attack. A study published in 2018 exposed erectile dysfunction as an independent risk factor for cardiovascular diseases. The study (Multi-Ethnic Study of Atherosclerosis; MESA) was mainly concerned with the risk of atherosclerosis, and over 6000 men and women were the subjects. Of the men, 1914 were interviewed about possible potency disorders; 877 had erectile dysfunction – their average age was sixty-nine. Over an observation period of four years, the risk of a stroke or heart attack was two and a half times higher for the men with potency disorders.[14] It seems to gravely affect the hearts of men when their penises fail. And vice versa. They obviously become sick. Repercussions of heart diseases and circulatory disorders on women's sexuality are unfortunately much less researched. Moreover, heart disease in women continues to be more rarely diagnosed than in men, which is also due to the different symptoms. To what extent risk factors for the heart also include risk factors for the functioning of female sexual organs can, at the moment, not be said with any certainty. However, illnesses such as diabetes or severe obesity appear to also affect women's sexual health.[15] The question of whether reduced blood supply to the sexual organs constitutes an independent risk factor for apparent cardiovascular diseases in women, as it does in men, cannot be answered at present – although there are hints that it might indeed be the case.

We smell with the heart

Smell influences our choice of partners, and our nose likes to join our amorous play. But not just the nose! In 2017, receptors were discovered with which the heart can smell.[16] However, it doesn't smell 'raspberry' or 'cedar', but fats in the blood. And when they do not agree with it, it will slow down and beat less powerfully! When I was at university, if someone had speculated that the heart could smell, they would not have been taken seriously. The scientists who have now discovered the heart's nose suggest that scent molecules may also get into the bloodstream via the skin and then be perceived by the heart.[17] For millennia we have known that essential oils can calm or revive our senses. A few drops of oil on the chest, rubbed in with loving movements: a little massage can be a balm and open lovers' hearts. And not only their hearts …

It is sometimes already visible in the eyes. If they are wide and the cheeks red, this can be a revealing and promising sign. Then adrenaline, which also comes from the heart, will get to work. To imitate this 'look of love', Venetian women used to swallow the juice of the poisonous *Atropa belladonna.* This supposedly rendered them irresistible to men, and that may be a reason for the botanical name of deadly nightshade: *bella donna* means 'beautiful woman'. People's longing to help love along with potions is age-old, and in honour of the goddess of love, Aphrodite, they are called aphrodisiacs.

Many aphrodisiacs have been examined predominantly for their neurobiological effect on the brain. On the question of whether oysters, shark fins, ginseng or a specific 'chicken soup cooked with love only for you' will promote or hinder the release of dopamine, serotonin, oxytocin or endogenic opiates, the results are inconclusive. In an interesting scientific analysis on the pharmacology of love, the authors come to the conclusion that it is different for every couple. Too many unknown factors play a role. The authors contend that it is impossible as yet to anticipate the 'magic moment' of falling in love and then immediately put the subjects under a scanner.[18] I deeply respect my colleagues, but allow me one remark: maybe they are looking in the wrong spot. For if hitherto unknown receptors on the heart were only discovered in 2017, that shows that the final chapter in heart research is far from written.

THE LONELY HEART

Kordula was a petite woman in her mid-thirties, with an alert mind and dark curls. She ran a successful business and was a caring and loving mother to her two small children. She combined work and family seemingly effortlessly: her husband, an engineer, supported her in both fields. Kordula and Achim greatly valued self-fulfilment and freedom. She did not mind that he quite frequently went on climbing trips for several days with his mountaineering friends. She was more the ocean type, preferring scuba-diving trips and walks on the beach. To please Achim, she had tried climbing, but remained afraid of it.

Even though the children would have liked to do something with their father on the weekends, he increasingly preferred to make conquests – and not just in the mountains. Kordula had a hunch, and in her heart sometimes felt a tightness, a heaviness, a pulling. When her hunch turned out to be true and she found proof of Achim having affairs, she was immeasurably disappointed and furious. But as a business leader she had learned to control her emotions, especially when she had a specific goal. And she had one: she wanted to save her family. After careful consideration she decided to continue with her marriage, even though it would be mostly in the interests of her children. Surely her husband would change one day, it was only a phase – and until then she would accept his escapades. She made a reasonable decision, as she was used to doing in her work – a decision in favour of the project 'family'. And then she got back to her everyday life and made an effort to pretend nothing had happened. Achim, too, made an effort, but Kordula doubted he was faithful to her. He was probably just more careful. At one stage she thought about employing a private detective. But that would have been very expensive. So she decided to remain true to her decision to keep her family together and look the other way when necessary. It would surely end at some point. Achim would come to his senses; despite it all he loved his children more than anything and would never jeopardise his relationship with them.

After a few weeks, Kordula's heart began to palpitate, at first only at night. Then it would beat in her throat and hammer irregularly in her chest. She slept badly, sometimes for only a few hours. But she was disciplined and stayed in bed. Her body had to recover, as she had to get through long to-do lists every day. Her strategy seemed right, as the palpitation

horter. But then her heart started
ng the day too. This began during
was beating so hard Kordula had
ld not accept this anymore. She
at the firm. That same evening,
was a doctor. A little bit. She
lessness. She kept the trigger,
up in her heart. Her ECG, her
s were normal, and the doctor
iques and more rest periods
s assiduously. She listened to
to sleep, and that seemed to
curred less often.

, when her heart did not
though everything was
d hidden Easter baskets
ey laughed and tussled
ren's voices mingled with
bell clapper struck her
nt than ever before, and
e to calm down, she com-
k.

he replied. But that was
ishly wondered what she
Achim wanted to come.

'Do everything as planned,' she tolc
been looking forward to it. Go to th
I'll be back soon.'

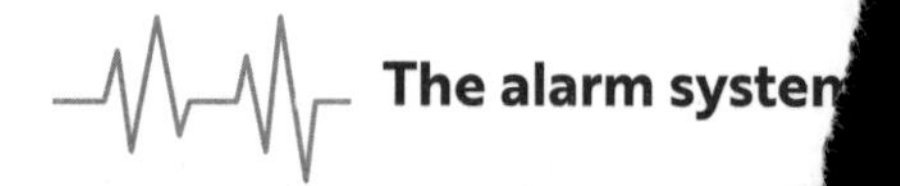

The alarm systen

In the clinic, she was diagnose
this kind of arrhythmia, the t
with the rest of the heart: the
blood into the ventricles. The
by 15 per cent. Patients no lo
of breath quickly. Often th
beats like crazy, as was the c
that strong emotional str
whelmed quite frequentl
our needs over a longer p
a balance, the heart will
consequences may be se
clots will occur which ma

In the clinic, Kordula wa
the heart valves – or occ
can also cause atrial fibr
neither the ultrasound
showed changes in the h
valves and the size of th

started to abate or get s
making itself known duri
an important meeting. It
to leave the room. She cou
had to be fully operational
she confided in a friend who
spoke about stress and sleep
Achim's unfaithfulness, locked
blood pressure and other result
recommended relaxation techn
during the day. She kept to thi
meditation music before going
work: the racing of her heart occ

This worked until Easter Monday
want to calm down anymore, even
perfect. Together with Achim she ha
for the children in the garden. Th
while they searched. The light child
the ringing church bells. A mighty
heart, whose beats were more viole
she had trouble breathing. *You hav*
manded herself, but it did not wor

'Mama, what's wrong with you?

'It will be better in a second,'
a lie, and she knew it. She fever
should do. Then she called a taxi.

l him. 'The children have
e zoo without me. I'm sure

n

ed with atrial fibrillation. With
wo atria no longer beat in sync
y flutter and no longer transport
heart's performance decreases
nger feel productive and get out
e heart also starts to race and
ase with Kordula. Today we know
ain, fear and being generally over-
y lead to arrhythmia.[1] If we neglect
eriod and do not succeed in finding
at some point sound the alarm. The
vere – in the fluttering atria, blood
ay cause a sudden stroke.

s examined thoroughly. Illnesses of
asionally even too much alcohol –
illation or high blood pressure. But
of her heart nor any other tests
eart to match the symptoms. Her
e heart chambers, her blood test

results and her inner organs – all appeared normal with Kordula. And when someone asked her how she was, she always answered truthfully. But she remained on the surface, describing only the racing of her heart, forgetting her hurts, the disappointment, anger and worries. She was in a hospital, after all. Apart from that, wasn't it the best strategy not to pay any attention to Achim's infidelities? She did not want to give them too much weight. And surely the tablets would help to normalise the rhythm of her heart. But they didn't. Only their side effects hit Kordula really hard. She felt more fatigued, and her general condition was abysmal.

There are many drugs for arrhythmia, but nearly all of them share one feature: they block certain channels to the heart, for example for ions such as sodium and potassium, or they block heart receptors and thus the heart's ability to sense bodily states. In a healthy heart, everything is in flux and in a very fine balance. It is sometimes possible to force the heartbeat back into its proper rhythm by blocking certain feelers, for example with beta blockers. Millions of people take beta blockers or other such substances daily, often for decades. But they not only affect the heart, but the whole body, and can lead to fatigue, listlessness, bad moods due to depression, memory disturbance and loss of libido.

The doctors decided to administer an electric shock to Kordula's heart so that it would finally behave once again.

Under anaesthetic, she was connected to various heart monitors. Electrodes the size of a palm were stuck on the right and left sides of her thorax and connected with strong cables to a defibrillator. To be able to see exactly what was going on in the heart, an ultrasonic probe was placed very close to her heart via a tube through the oesophagus. And she was tied to the table.

It is important that the electric shock is synchronised and is fired at a point during the heart excitation when it does not produce a fatal fluttering of the ventricles but the correct rhythm. The software of the device takes care of that. The red button with the lightning symbol was pressed, and Kordula's body twitched violently, her face grimacing, her head jerking. The monitor showed a flat line.

I know this moment very well; during many heart operations or afterwards the heart needs to be defibrillated. After the electric shock it is still for a second or two, shocked in the true sense of the word. Until the heart catches itself again and the new rhythm sets in. I have often experienced this stillness and realised with astonishment, after the defibrillation, how long one second can feel. Sometimes it seemed to me as if my own heart had stood still as well.

Unfortunately, the rhythm of Kordula's heart did not change, even after two further attempts with increased voltage. Only her body twitched ever more violently while the heart remained unfazed. And so she was discharged from hospital, as often happens with arrhythmia, or more specifically atrial fibrillation with an unknown cause. Ironically, this is also called 'lonely fibrillation' as no illness has been found in the heart. In addition to the tablets to regulate her heart's rhythm, Kordula was also given blood-thinning drugs to avert the danger of a stroke.

When she was back at home, she felt as if she had aged many years, and felt very, very ill. It seemed to her as if the world had lost its colour. Suddenly everything was only black and white. Melancholy thoughts afflicted her and she perceived many things she would usually do effortlessly as an insurmountable struggle. She did not recognise herself. Her family doctor prescribed a mild antidepressant. It worked in that she cared even less about things. What mattered was that she was able to function in her business and at home. With incredible self-discipline she managed to look after the children and lead her business. At 9 pm she would fall into bed exhausted, while Achim would often stay awake until after midnight, playing poker on the computer – his new hobby, he said. Kordula did not even want to know. She felt deep inside that something had lost its beat in a fundamental way.

Then she heard about a method that had been successful for 50 to 70 per cent of patients with similar symptoms, and full of hope she made an appointment at a heart centre.[2] There she was advised to undergo a procedure called catheter ablation. Using local anaesthetics and fluoroscopic imaging, a catheter would be pushed, through the blood vessel in the groin, to the heart. With this method, cardiologists disable faulty electrical signals in the lung veins close to the heart and in the atria; circulating excitations are interrupted. However, the heart can also be badly damaged by such a procedure. For example, the flow of blood from the lungs may be hindered, or in rare cases life-threatening bleeding may occur. Kordula would possibly need a pacemaker, as the subtle control and permanent adjustment of the heart speed via the nervous system and the communication with the brain and the other organs may of course also be impeded.[3] Imagine it as if we had cut a telephone cable. The patient would have more peace but would also not be able to receive calls anymore.

'I'll think about it,' she said.

Achim did too. He was greatly worried, and a voice inside him (maybe coming from his heart) told him he was not completely free of blame for his wife's arrhythmia. He wanted to change and assured Kordula of his support. Those were well-intentioned words, and she believed him, but her heartbeat remained unfazed.

In the days that followed, she talked to many people about the imminent decision – and that is how I got to know her. A friend told me about her case and asked me to advise her over the phone. She wanted my opinion.

'You know all about the heart. What should I do?'

In the course of our conversation I found out that she was placing her highest hopes at the moment on homeopathic therapy. An experienced GP, who had been recommended to her, had long been devoted to homeopathy and wanted to try this path. Dr Herbst told me so herself when I spoke with her on the phone. She had all of Kordula's results. She seemed reputable and competent to me; I also liked the way she spoke. 'What heals is right,' she said quite a few times.

During the consultation, Dr Herbst had listened to Kordula for a long time and asked a lot of questions no other doctor had asked before. Among other things, she inquired if something in her life had recently changed dramatically. Kordula burst into tears and spoke of her husband's infidelities.

Diagnosis of atrial fibrillation is as common as that of depression, and the risk increases with age. Often both illnesses occur together. They are nowadays both considered to be widespread. At an international heart congress in 2018, there was intense discussion about how closely the health of heart and brain are connected. It is not only the case that atrial fibrillation can lead to depression – a very recent study proves that it can also be the other way around: in patients with depression the risk of atrial fibrillation rose by about 30 per cent.[4] Dr Herbst told me that she was convinced she would be able to help Kordula. 'I am sure that my remedy will hit home. I am confident, dear colleague, that you will not have to operate on this heart.'

Years ago I probably would have shaken my head at such a claim. But I had learned a lot recently. 'What heals is right,' the doctor said again, as she bid me farewell.

To heal the beat

In my opinion, the word 'healing' is a big word and therefore rather rarely used in medicine. Generally it is understood to mean the restoration of someone's physical and mental health, or at least a restoration to the initial condition before their illness. To heal hearts was my profession, but I left considerable scars, and an operation can also have serious side effects – so the term 'repair' might possibly be more accurate. Sometimes I also repair the rhythm of a heart, as atrial fibrillation doesn't always occur by itself but can happen together with a defective heart valve, which has to be repaired in openheart surgery. In such cases I also try to get the badly disturbed beat of the heart 'back on track', as in the case of Mr Laschek.

I looked into his open chest and directly into his heart. It lay motionless before me; the heart-lung machine was softly humming. I had first repaired his leaky mitral valve. Of all the valves, I like this one the best, because its two valvular cusps give it the appearance of a smiley when they come together with their half-moon shapes. When the cusps of the mitral valve can no longer properly direct the blood flow because their geometry has changed, the valve becomes leaky and its smile vanishes. With several cuts and stitches I had given the cusps a new shape. Now a ring was also implanted, and the face got an outer frame which gave it extra stability. Of course this isn't plastic surgery but a vital intervention – but the mitral valve was looking friendly again. To check if it was still leaking, I filled the left ventricle with a sterile saline solution and observed if it would flow back to the atrium. No leaks. Good. Now I could take care of the atrial fibrillation by ablating certain nerve pathways. When doing this I always felt like an electrician, as with atrial fibrillation the heart produces something like short circuits which backfire. I could emit high-frequency radio waves at the push of a button with a device that looked like a pen. With the resulting heat, it is possible to sever nerve tracts in a subtle way. I could not see them as they are very delicate and hidden in the heart wall. But in my mind's eye I had a map of the heart with all nerve tracts marked. I put my heart pen on the spots where I visualised the tracts. It looked a bit as if I was scratching signs into an old tree. I wished the truth about the heart would trickle out like resin, but instead I only left scars which electrically separated the heart from the lung (pulmonary) veins. Then I sutured the 'heart ears' (atrial auricles). Yes, the heart has ears. That is what the excrescences of the atria are called which look like two big ears. They are what makes the heart heart-shaped, and in them the ANP (atrial natriuretic peptide) is produced. During the operation concerning atrial fibrillation the heart ears are closed off from the inside with stitches, because otherwise dangerous blood clots may occur. Sometimes they are simply cut off because we believe that a large amount of the nerve tracts involved in the atrial fibrillation run inside the ears. I feel sorry about that every time as I like the heart ears. Maybe we will one day find that we hear more with them than we can prove today. Who knows, perhaps that was what the old anatomists were thinking, too, when they gave the heart ears their name. After I had closed them off, I also sutured the atrium and opened the blood flow to the heart. Mr Laschek's heart started to beat again after a few seconds, and to my great delight it was pumping with the right beat.

Most illnesses of the heart lead sooner or later to arrhythmia. Sometimes the brain is involved, too, with stress, fear or depression. Whatever the causes, at the end comes a disruption of the energy flow through the heart. It can be treated in different ways, from the finest energies of homeopathy to high-frequency surgical isolation. Both methods treat the heart's energetic disturbances with energy. To treat like with like is also the fundamental principle of homeopathy. Perhaps the medical worlds of healing the heart are not as far apart as it appears at first glance. 'If your only tool is a hammer then every problem looks like a nail,' said the well-known psychotherapist and scientist Paul Watzlawick. That is why I, as a heart surgeon, deem it important to be familiar with a big repertoire of tools for the heart. Sure, only surgery can help with a gravely ill, anatomically deformed heart valve. However, when arrhythmia is of a mental or psychological kind, it makes less sense to tackle it at the heart. In such cases we can work with more subtle energies which involve the whole person. We should have the courage to lose the blinkers: what heals is right!

Four months later, I heard that Kordula's arrhythmia had vanished. I was curious, called Dr Herbst and inquired about the name of the homeopathic remedy she had prescribed. It was a high-potency mineral salt, strongly diluted and then exponentiated by multiple shaking. Homeopathy acts on the assumption that the dissolved matter becomes immaterial, energetic information. For many people that is incomprehensible. For this question of faith, a professor of experimental surgery always comes to my mind, who gave me the following advice in my last year of studies: 'You can't be a good surgeon if you haven't understood the nature of matter. Inside you, matter is energy, and both can merge into each other. If the surgeon who you want to be one day, a surgeon to whom patients entrust their bodies, has not grasped that, he has also not understood the nature of wounds and healing.'

In the black hole

I am glad whenever I have time for my Porsche's six-cylinder boxer engine. I love its roar; its heart is already a little older and air-cooled. On this mild summer evening I drove not only through the beautiful landscape near the Baltic Sea, but also through my own past. I could even hear the voice of my former professor. Back then he was a pioneer in his field who had devoted himself to the development of new surgical techniques. In his free time, he organised homeopathic seminars, held in a small lecture hall. These events were not printed in any university calendar. It was a quaint bunch which gathered late in the evening, as if for a conspiratorial meeting. There were country doctors, veterinarians, students, philosophers and professors from different fields. I especially remember a biochemist with long white hair and a big bushy beard who found homeopathy downright 'logical' and also tried to enthuse us all about quantum mechanics.

After a few months of attending this circle I lost interest. I wanted to become a 'proper' surgeon, to cut and stitch, to lay bypasses and replace valves. But had I understood the nature of matter and the energy of the heart? Of course I was familiar with the most famous equation in the world, Einstein's theory of relativity: $E = mc^2$. But what was behind it? According to the theory of relativity, little amounts of mass can become large amounts of energy, and vice versa. All you need to do this is the square of the speed of light. A speed which the heart of my Porsche was not capable of. Unhurriedly 'we' were roaring home. In the evening I became lost in literature. It seemed to me as if I was comprehending some things in a new way. I was no longer the student, the young doctor from back then. I was a heart surgeon in search of the true heart. Matter and energy are the essence of all being, inextricably connected, from the smallest parts of atoms to gigantic galaxy clusters.[5]

That night I sat over my books for a long time, surfed the internet, and when I finally fell asleep had wild dreams. At the speed of light I rode on atoms through galaxies that were expanding further and further and were eventually gobbled up by black holes. In these black holes, light was kept prisoner, and so was I. In the morning I woke up soaked in sweat and with a madly beating heart. I was confused. Where was all this going to lead me?

Some weeks later I became acquainted with the British sense of humour: subtle eccentricity and exquisite learnedness. Two world-renowned Oxford professors, heart researchers Denis Noble and David Paterson, converse about the future of heart medicine in a series of video interviews.[6] Paterson shares my view that the greatest challenge for modern medicine is to bring back together the fragments of the human being (examined in every genetic detail) and integrate them in holistic medicine. This, he said, is the task of an interdisciplinary, transnational medicine – at the intersection of mathematics, physics, chemistry, information technology, engineering, social sciences, psychology and philosophy. These intelligent words impressed me enormously, and I decided to further extend my search for the secrets of the heart and to think 'bigger' than I had been trained to do.

During my studies and in many places still today, cardiologists taught about the heart, neurologists about the brain and psychiatrists about the soul. This reductionism of modern medicine has disassembled the human being into organ disciplines and molecules. On the one hand, that makes sense in order to understand functions in detail. On the other hand, the connections and important interrelations are neglected. The whole is, after all, more than the sum of the parts, and this 'more' lies in the complex interactions of organ systems, body, psyche and of course the environment. But which share could the heart have in the whole? Even though we sense them within us, we have so far not found any sensors in the heart for compassion and love. If you look very closely, the lovely German word *wahrnehmen* (to perceive) contains a very meaningful first part, namely *wahr* (true). So far I had never asked myself this question – is what we perceive (*wahrnehmen*) with our senses, what we hear, see, smell, taste, touch and feel, actually the truth?

THE BIG PICTURE

In the philosophical sense, everything that exists is true – and real. This may be considerably more than what we can objectively perceive in a physiologically measurable way. And it can far exceed our imagination. Actuality is significantly different, in terminology and content, from reality, which is derived from the Latin word *res*, meaning 'the thing' or 'the matter'. Thus, reality is material. We call 'real' those things we can perceive with the classical five senses. Objects that we can touch with hands and skin and see with our eyes. Food we can smell with our nose and taste with our tongue. And of course noises, which we hear with our ears. The sensory organ that is the eye allows us to see certain wavelengths of light, which manifest themselves as colours in our inner representation of reality. What we can perceive with our senses is not truth but reality – which is only a highly selective excerpt of the truth (which is the existence of everything).

There are considerably more wavelengths of light that we cannot see: for example, X-rays or radio waves. We can hear certain soundwaves, but some animals see, hear or smell much more or in an entirely different way than us human beings. Migratory birds are guided to Africa by the Earth's magnetic field, which is inaccessible to our senses. Whales, mammals like us, are guided through the vast oceans by it. Bats possess an echo system: radar signals are their reality, and a black image with lots of white spots (reflections of mosquitos) means a well-fed night for them. Dogs smell those sausages long before we take them out of our shopping bag. Their highly sensitive sense of smell shapes their reality differently as they live predominantly in a world of odours.

We know all that, and in our scientific search for truth we expand reality to include phenomena we can find proof of, even though we can't perceive them with our five senses. For example, for a long time there was scientific consensus that we are made up of small particles, namely atoms. The model of atoms and their interactions and connections to form molecules allows us to explain our real world quite well, even though nobody has ever seen a singular atom. 'When such a model is successful at explaining events, we tend to attribute to it, and to the elements and concepts that constitute it, the quality of reality or absolute truth,' wrote Stephen Hawking and Leonard Mlodinow.[1]

The model of atomic reality was replaced nearly 100 years ago by quantum mechanics, which asserts that everything is made up from elementary particles. However, one cannot imagine elementary particles to be like mini Lego bricks or atoms. They are much smaller and lead a hybrid existence, as they can be both particles and light waves. Elementary particles can be in different places, millions of light-years apart, at the same time, and in many ways they behave 'spookily', as Albert Einstein called it. They like to synchronise themselves like pendulum clocks do, but curiously they seem to notice when they are being watched. Then they determine if they are a particle or a wave. How macroscopic matter, including human beings, is created from this strange mix – that remains the topic of many models of reality, theories and discussions.

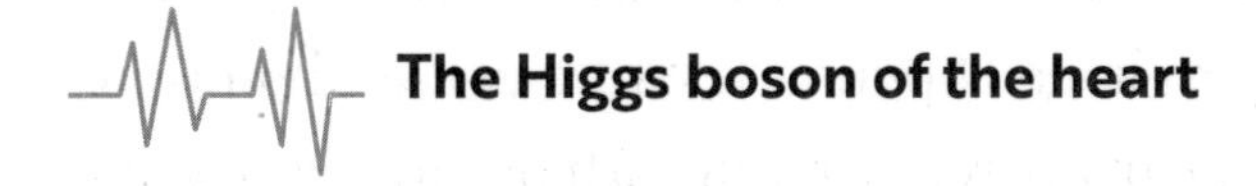

The Higgs boson of the heart

What is needed, in any case, is the Higgs boson. In July 2012, the existence of the Higgs boson, also called the 'God particle', was confirmed in CERN's Large Hadron Collider. British physicist Peter Higgs had already suspected in 1964 that the particle had to exist, as otherwise quantum mechanics would have a hole. This was met with a lot of scepticism, and at first no one wanted to publish this theory. In 2013, Higgs was awarded the Nobel Prize in Physics. For physicists, the Higgs boson is a heavyweight as it is the particle, or rather a field, which had been missing from the modular design of particle physics and which gives mass to all other elementary particles.

In quantum physics, everything that exists is connected and in constant interaction, constant exchange. That is why quantum physicist and winner of the 1987 Right Livelihood Award (also called the Alternative Nobel Prize), Hans-Peter Dürr argues: 'Basically, there is no matter. At least not in the familiar sense. There is only a structure of relationships, constant change, aliveness. We find this hard to imagine. Primarily, only correlation exists, connections not based on matter. We could also call it spirit. Something which we can only experience spontaneously but not seize. Matter and energy are secondary features – as congealed, solidified spirit, so to speak. According to Albert Einstein, matter is merely a diluted form of energy. Matter's basis, however, is not some refined kind of energy, but something else completely, namely aliveness.'[2]

Are these insights really completely outside our faculties of imagination and perception? Or do we feel intuitively that they could be right? One day I had the thought that the heart, too, is a Higgs boson. It gives all other particles life. Or could there be more Higgs-like entities in the heart? A field that gives love and truthfulness and all the other qualities we intuitively ascribe to the heart? The Higgs boson of the heart could be an elementary module of the particle physics of humanity.

The Big Bang of the heart

At the other end of the scale from the particles that are even smaller than atoms (quanta) sits the universe – and we accept as reality that it was created with the Big Bang 13.7 billion years ago, and that we are in a wider sense made of stardust, of elementary particles.

Before the Big Bang, everything was one, all existence compressed in a tiny spot. This state is also known as the singularity. Ever since the Big Bang, the universe has been expanding. The most well-known theory in this regard is Einstein's theory of general relativity, which deals with space, time and gravitation. There are also theories according to which the Big Bang was not the beginning of everything and there had been a universe already before the Big Bang which had contracted to a singular point and is now expanding again. For example, the theory of 'loop quantum gravity' by German physicist Martin Bojowald views the Big Bang merely an expansion phase of the universe, between periods of contraction.[3] In his theory there are not only the smallest particles of which matter consists, but also the smallest indivisible space-time atoms. In summary, and very simplified, they form a space-time mesh which expands and contracts again. Like an eternal heartbeat of the universe. Our small human heart, which does its *ba-boom* for a lifetime, would thus have a big brother: the Big Bang. All that exists follows an eternally oscillating rhythm of exertion and relaxation, the beat of the heart and of life.

Soap bubbles?

In the world of the smallest of the small, of elementary particles and quantum physics, gravitation does not feature. In the two theories of the Big Bang (the general theory of relativity and quantum mechanics), it either plays a very prominent role or no role at all. That is to say, they 'are known to be inconsistent with each other – they cannot both be correct', as Stephen Hawking explained in his book *A Brief History of Time*.[4]

If we calculate back to the time of the Big Bang, we reach a point where the theory of relativity and all physical laws lose their validity. In other words, we have two models or theories of actuality which explain a lot of things very well but in the end contradict each other. And thus current physics applies a lot of energy to the development of a theory which conciliates the theory of relativity and the laws of quantum mechanics and might also be valid when discussing the Big Bang or the time before it.

The best-known attempt to bring together the theory of relativity and quantum physics is string theory. According to it, elementary particles are no longer particles but one-dimensional strings of energy that can also form matter. Based on this, an infinite number of universes can exist. This model is called multiverse theory (M-theory). If the universe is already infinite, the multiverse has to be even more infinite. We are unable to imagine this. It is theoretically possible that our universe exists not only beside but also within other universes – like a soap bubble within another. Or it might be that other universes are located within our own, and that parallel worlds exist. The whole thing – or the truth, we just don't know – would unfold (according to the M-theory) in an eleven-dimensional space.[5]

Biology and medicine are concerned with matter and the nature of all living things. Their models of objective truths are still based on atoms and molecules. With them one can explain many so-called metabolic processes which relate to what is substantial, material, tangible inside us. But the heart poses riddles here, too, especially concerning the cells from which it is made. The growth behaviour of these cardinal stem cells generates different results in different laboratories. The idea of uniting such discrepant findings in a theory, a string theory for heart cells – analogous to the string theory for the universe – was introduced in 2015 in *Circulation Research*, a renowned scientific heart journal.[6]

Molecular biology and genetics are viewed as the postmodern ultima ratio of medical research. This is enormously complicated, and our human minds would hardly be able to grasp the complexity of a quantum-physical consideration of the human organism and its connections with everything that exists – even though such an account could possibly get close to reality. Hans-Peter Dürr put it concisely in one of his lectures: 'The new reality is completely different from what we imagined it to be.'[7]

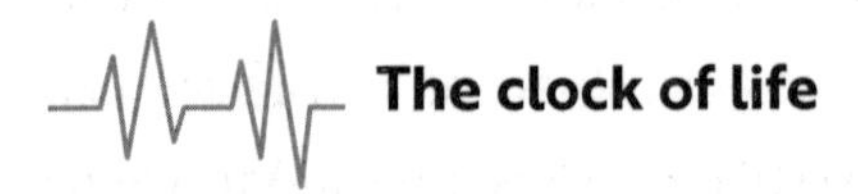

The clock of life

Even though we are not material in our innermost being, the elementary particles our body is made up of join to form the material substances that constitute our body. The body is that which we are, that which breathes, speaks, feels, thinks and loves. Similar to us humans, Earth, too, consists in its innermost regions of glowing energy which at the surface becomes a living organ of earth, fire, water and air. Bacteria and plants live on it and work the magic of using the light quanta of sunlight to put biological matter and oxygen into the air. This process is called photosynthesis; many of us learned about it at school. Today we know that both photosynthesis and the transformation of sun energy into the salad we eat work according to quantum mechanics. Like our Earth, every red blood cell has inside it a core of iron, only it is no longer glowing. This single iron atom binds oxygen, and then the blood in our lungs changes its colour from dark-red to light-red, and that means life. As far as we know, Earth is the only place in the universe where such things are happening. However, we do not know where exactly it is located. In a space that is infinite in all directions, we cannot calculate our absolute position. We only know the distance to our neighbouring planets and the orbits on which these planets move. What we call time came about in the same way. The Earth revolves around itself, the Moon revolves around the Earth and both together revolve around the Sun. The side of the Earth that faces the Sun changes with the revolution. If you stand in one spot on the Earth, there is a moment when the sun rises and one when it sets in the evening. We call this phenomenon – the sum of the moments (one could also say heartbeats) between sunrise and sunset, the alternation of light and darkness – the 'day'. Scholars of astronomy and physics broke it down into units of time. Time, however, is not an existing, measurable quality like amperage, magnetism, gravity or temperature. Its existence is therefore deemed an illusion by many physicists and philosophers. When we observe how things change, we say that it is happening within a certain time. In physics, time was introduced as a quality that served to describe the duration, speed and sequence of events. For example: it becomes day and then night, a flower blooms and withers, we are born and die. However, day and night are relative concepts of time and a question of location. When I was in New Zealand, my son Josef was having breakfast when I went to bed. It became morning for him when evening was approaching for me.

What do we learn from this? We do not know our absolute coordinates within the universe. We are located somewhere in infinity. Our navigation device shows us merely the neighbouring stars and galaxies. In other words: we do not know where we are, or what time it is.

But you can put your hand on your heart and feel your heartbeat and your breath. The heart is beating its life-long rhythm of the moment. You can feel it with every heartbeat: now ... now ... now. Every now is the present. Your present. The heartbeat is the clock of life, your clock of the universe, from moment to moment. You can only experience the present. The past is past, and the future not yet here. You cannot deliberately stop this clock. You can only calm down, breathe and let your heartbeat and thereby your life become slower, thus creating your very own experience of relativity.

The universe contracts and expands, there is inhalation and exhalation, day and night, life and death. Every heartbeat is a process of creation and creates life immediately. Every time the heart relaxes, it fills with blood, and every time it contracts this blood is pumped through the body. This is a vital process, and if it is interrupted for only a few minutes, biological death will be the consequence.

Then our heart and our time will stand still.

The only real time is the heartbeat. When my *ba-boom* ceases, my time has elapsed, at least for me as an organic being.

HEART CONSCIOUSNESS

Some people are able to feel their heartbeat consciously and even count it precisely inside themselves. Scientists have a keen interest in them, because the heart rate is a measurable, verifiable number, not an intangible feeling. From studying those who can intimately *feel* their heart rate, researchers hope to gain new insights about the consciousness of the heart – and they report interesting findings. In one study, healthy students were shown different movie scenes. Good 'heartbeat counters' felt emotions such as fear and anger (but also joy) considerably more intensely than those who were less well able to feel their heartbeat.[1]

But the conscious perception of our heart intensifies not only our emotions but also our empathy with others. Students who were more able to feel the frequency of their heartbeat could also more accurately distinguish if faces in pictures were happy or sad.[2] A study published in 2017 describes that those who can feel their heartbeat more accurately behave less selfishly in money affairs and don't only consider their own interests.[3] Even though the financial generosity was only measured in a game, the data shows that the good heartbeat counters display heightened sensitivity towards themselves and others. On the other hand, the sensitive heart perceivers also become stressed more easily. In exam situations they performed worse and were plagued by negative emotions.

Should you not feel your heartbeat at the moment, you have no reason to be alarmed. You just belong to the majority (65 per cent) of people who do not feel their heartbeat when at rest, and I assume you are reading in a relaxed state, not while you are jogging. Numerous factors can play a role in heart perception when one is at rest; it depends on one's physique as well. A common mechanism of heart perception is the beating of the heart against the left inside of the chest. This so-called apex beat is felt more frequently by slim people, men, and athletes with large hearts. The pulsing of our arteries inside us – in the head, stomach or throat – are other possible ways through which we can perceive our heart's rhythm. However, some people manage to do it just like that, without any pulsation, merely via nerve signals or because they have a very pronounced sense of their bodies.[4] I don't believe that this perception makes them better people. But I recognise in the work of current consciousness research an increasing interest in the heart as a place where more may be found than was so far assumed.

Panic because of the pump

Not all people are glad about being able to feel their heartbeat. Many suffer when they frequently feel their heart intensely, because it worries them a great deal. Especially when they not only strongly feel the rhythm of their heart in general, but sometimes feel it to be irregular and connected with pain, chest tightness or shortness of breath. Often the diagnostic apparatuses will not show any abnormal results. If the panic because of the pump continues, doctors speak of cardiophobia attacks or cardiac neurosis. The causes are often deeply hidden. As a surgeon I know that wounds heal better when they are looked after. Daily dressing changes and wound examinations are compulsory in every surgery department. Wounds of the soul, too, have to be seen in order to heal. If we continuously ignore them, the heart will make itself felt at some stage.

I find that the voice of the heart is really loud in our childhood. It knows what makes us happy, it knows what makes us sad, it knows what makes us strong. The hearts of young people are still very sensitive and malleable. This voice becomes quieter the more we think, the more we are educated and shaped, the more our heart is hurt. Then one day some of us do not hear the voice at all anymore. The mere memory of the injuries to our heart continues to hurt a lot, and many of us therefore do not want to feel their hearts anymore – after all, feeling always means feeling pain, doesn't it? Thus, they become ever more insensitive in matters of the heart. I have met more than a few patients who were even ashamed of the injuries to their heart and did not want to disclose them under any circumstances. Over the years they lose access to themselves, and the formerly protective armour becomes hard and tight and rigid. But the voice of the heart is still there. It lives as long as we do. However, if it is continuously ignored, our heart will one day start to beat and race in a way that will greatly worry us. And we become afraid because we have long forgotten the cause. The time of injury is not always as recent as with Kordula, whose husband cheated on her. Sedatives and beta blockers are no long-term solution, but make us heart-deaf. It is better to lend an ear to this voice finally and start at the heart to retrace the path to the root of the wound. This is best done in the company of an experienced doctor who not only knows the soul very well but also the heart. Such patients need courage to engage with their almost-forgotten heart, trust to follow its voice, and finally compassion and love for themselves.

Astonishing research shows how the pure mechanics of the heart, with its contraction and relaxation, influence our perception profoundly via nerve tracts. Faces are perceived more intensely and threateningly when subjects see them during the contraction phase of the heart and the feelers for the pressures in the ventricles are activated. Simultaneously the amygdala, our centre for fear and anger in the brain, becomes active and is supplied with more blood. Images are deemed less threatening when the heart is relaxed and filled with blood.[5] Because our heart contracts and relaxes permanently (depending on the strain, this process is sometimes stronger, sometimes weaker), what is important is the net effect: the balance of exertion and relaxation across a longer period of time. If we perceive the world as friendly or hostile also depends on the pump function of our heart, the pressures in its cavities and its signals to the brain. For this reason alone it is important to be balanced in the heart and to feel one's voice. In a twist on the old saying: Man proposes, but the heart disposes.

Cardio-cognitive consciousness (are you still beating or already feeling?)

Regarding the fundamentals of the development of consciousness, scientists take their point of departure at the cerebrum and brain stem and then gradually slide lower, to the organs. That is where its roots are, and it grows towards the brain from there. I want to explain this using simple examples: when we are hungry, this is a feeling which describes a specific bodily state – a falling blood sugar level. When our heart is not properly supplied with blood or its cells are inflamed, pain may occur there. Pain is also a feeling and informs us of the fact that our body is being damaged. Further examples for simple feelings are the sensation of thirst or warmth, or our breathing. A feeling is thus, first of all, a vital sensation of our body. It informs us how the body is and what it needs. To feel one's own body is the simplest, most basic form of consciousness. There is no clear definition for it anyway, but consciousness does not mean to be intelligent or creative. It is the perception of one's own bodily existence.[6] And that, in turn, is not possible without the heart.

What we can feel with our hearts, however, goes far beyond the perception of one's own heartbeat. Sometimes when I follow a promising scientific lead and it seems logical to me, my heart goes into resonance, too. I feel in my heart that something makes sense. Or not, as the case may be, even though my arguments may be nothing short of impressive. If I see my children, I feel my heart overflow with love. For me these are not mere feelings, but complex qualities of heart consciousness – something I call 'cardio-cognitive consciousness'. And as we all, every now and then, perceive 'something' in the heart, these sensations would have to have their organic origin there. But how does heart consciousness get into the heart? Would it even be biologically conceivable that this special heart consciousness is transformed directly in the heart – meaning not in the brain but rather where we feel it?

What you seek is seeking you

During a holiday trip I strolled through Saigon's famous Ben Thanh Market and looked at the goods on offer at a butcher's stall. And suddenly I saw them. They lay next to each other in two bowls. The heart and the brain of a freshly slaughtered sheep, which are similar in their basic anatomy to those of a human. I stopped and looked at them for a long time.

I asked the shop assistant in English if he would let me touch them. He did not understand me. I showed him what I wanted. He gave a wide smile, and I saw that he had only two teeth left. One on top, one on the bottom. In this moment this seemed fitting since the recipes he recommended to me with lots of gestures didn't sound particularly easy to eat. Anyway, he allowed me to touch both the heart and the brain – or were they touching me? Cautiously I held them in my hands ...

There are moments in life when you suddenly see something you had not been able to recognise before. I looked at the brain in my right hand and the heart in my left. Both loved freedom above all else. A free heart and a free will are essential ingredients of life. These two organs which seem so different at first glance aren't so different after all! They are closely related. The words of the Persian poet and Sufi mystic Rumi came to mind: 'What you seek is seeking you.' If we seek the answer to a question with our whole heart, it will sometimes find us halfway.

'Ten dollars,' said the shop assistant. I would have given him 100. But what should I do with a brain and a heart in my hands? I bought a bag of spices from him instead and did not bargain when he said five dollars, even though it was daylight robbery. He had given me incomparably more. Astoundingly, heart and brain have numerous things in common.

Self-stimulation

Brain and heart are autonomous and can each stimulate themselves electrically. This self-stimulation occurs in certain ion channels in cell walls: the so-called funny current. They occur predominantly in the heart and the brain.[7] In the brain we find them in an area called the thalamus, which is presumed to be the gate to higher consciousness. But the heart, too, has its own pacemaker with which it stimulates itself for life, the so-called sinoatrial node. This node is made up of heart muscle cells, but they behave like nerve cells and have ion channels built into their walls for self-stimulation. Among other things, they take care that the heart beats continuously. It is assumed that the process of autonomous stimulation plays a role in the generation of our very own inner experience of thoughts and emotions.

Impulse conduction

Both heart and brain have their own impulse conduction system through which the stimulation spreads: in the brain via the manifold branchings of the nerve cells, which come together to form a three-dimensional net; in the heart via the heart muscle cells.

Heart muscle cells are something very special. In contrast to the muscle cells of the skeletal muscles, like nerve cells, they branch out to become a complex three-dimensional net whose fibres form the heart's organ structure with the chambers. Like the nerve cells, heart muscle cells can transmit electrical stimuli. This transmission works via so-called gap junctions – super-fast cell connections which also occur with the brain's nerve cells. In contrast to the synapses, which are much slower, transmission of stimuli is thus possible in both directions. The life-long adaptation of the brain, its connections and synapses to new demands is called neuroplasticity. The heart's nervous system and its musculature also adapt to new demands throughout life. Scientists speak of the muscular and neuronal remodelling of the heart.[8]

Control and communication

Both organs command both a complex biological perception function (sensor system) and motion control (motor function) and can themselves react to what they feel. This is called autonomous information processing and the heart, too, has its own nervous system to do this. In this way it can control its function in an extremely subtle way, but also continuously passes its data on to the brain, which can only perceive information from the blood in a very limited way because of the blood–brain barrier. One could say that the heart feels for the brain. They like to stimulate each other via nerve tracts, and we see the possibility for their communication also in the synthesis and distribution of identical hormones and the communication via biomagnetic fields.

Electromagnetic fields

The heart generates an electromagnetic field that is by far the strongest in the body; it is 100 times stronger than the brain's. Electromagnetic fields can store unlimited amounts of information. Without them, modern communications technology such as mobile phones and the internet would be unthinkable. Even before this, biomagnetic fields were involved in the communication between people, their brains and hearts; the electromagnetic waves of these organs can synchronise themselves and are in turn influenced by the Earth's magnetic field. In this way, communication between different creatures within an ecosystem is possible, and the heart is a decisive pacesetter.[9] My findings about the heart so far have shown me that it can send out messages of various kinds – for example hormones, pressure signals or nerve impulses – but it also has antennae to receive. I would be very surprised if that were not also the case for electromagnetic waves and if they would not allow us to receive 'heart information' (for example, to determine whether someone is being honest with us). We are also exposed to influences from the universe; a current long-term study, published in *Scientific Reports* in 2018, proves the influence of cosmic radiation on our autonomic nervous system and heart rate variability.[10] At the moment we can only speculate what specific meaning that might have for us.

Therapy

Neither organ works without electrical stimulation, which always has its own rhythmicity. Not only the heart is sometimes off-beat; the brain can display disturbances in its rhythm too. If the heart is continuously too slow, stimulation electrodes will be attached to its chambers and connected with a pacemaker implant. And what is good for the heart will often also help the brain. Not to help us think faster, but to calm down the brain's over-excitability during convulsive fits (epilepsy). For this purpose, the relaxation nerve, the vagus, is electrically activated with a small stimulator implanted in the neck. Thus, for the heart pacemakers hit the gas pedal, while for the brain they hit the brake, and both work with rather gentle electrical power.

That is not always enough – sometimes decidedly more electricity is needed. Nowadays AEDs (automated external defibrillators) are available in most public places as energy disturbances in the heart have become a rather widespread disease. You can recognise an AED by the lightning symbol inside a heart – and that is what they are for: they administer an electric shock to correct fatal arrhythmia. Even a layperson can save a life with this device. They just have to remember it's there! When a person suddenly collapses, call for help first. Then summon your courage and run or send someone else to get the AED. You don't have to be afraid; it will speak to you and do it all by itself. You only have to attach the electrodes. It is really simple! And you could save a human life. If you act swiftly, the probability of survival doubles.[11]

In the last few years another high-current method has been revived after being forgotten in the junk room of neuroscience for some time: electric shocks. Formerly they were used on convicted criminals and in psychiatry on the brain. It was hoped they would jolt the brain into functioning 'normally' again. Today electric shocks are applied to patients with severe depression. The method has been refined: the patients are under anaesthetic and the musculature is blocked by drugs to prevent the twitching of the whole body. Several studies testify to successes; how exactly the outcome is achieved, though, is far from being fully explained. What is certain, though, is that not only the heart but the brain, too, is being treated with electric shocks.

Packaging and networking

Mother Nature has packaged the two global players of our body subtly, protected them effectively and connected them extremely well. The brain is swimming in fluid: one could say in a tank of brain fluid encased by bone. This protects it from injury and gives it buoyancy. The heart is surrounded by the lungs and enveloped in bubble wrap, so to speak. Water and air are very permeable mediums for waves and particles, which connect us to our environment through quantum mechanics and energy. Sixty billion neutrinos per second rain through every square centimetre of our body. They come from faraway galaxies with black holes, from the sun, even from the Big Bang itself – and we are not conscious of them, don't notice them. Therefore it was assumed for a long time that these 'ghost particles' have no mass. But they do, and in 2015 the Nobel Prize was awarded for this discovery. The proof happened in gigantic water tanks with the purest crystal-clear water. In very rare cases a neutrino will hit a water atom core or electron and then 'blue lightning' will result from the electromagnetic radiation.[12] I ask myself what happens if a neutrino hits the fluid tanks around the brain or in the heart? Will the blue lightning be passed on via our 800,000 kilometres of nerve tracts or 100,000 kilometres of blood vessels? And what consequences would that have for our condition? In view of the sheer amount of these particles I think this is a legitimate question, and I hope that it can one day be answered.

Consciousness

If it is the case that electrical stimulation in the cerebrum's nerve cells is involved in the transformation of consciousness, then electrical stimulation in the space-creating, cardioplastic nets of the heart cells must also lead to elements of consciousness. There are numerous theories about how our experience of the self and the world can transform our body cells. Canadian biologist and anaesthetist Stuart Hameroff and well-known quantum physicist Roger Penrose assert that our brain's performance is based on quantum-mechanical processes[13] – in a subatomic reality (whose existence is no longer doubted by anyone today but which we cannot describe with our classical laws of nature) in which everything is possible, space is given to imagination and creativity, and poetry and logic are intertwined, like heart and brain. We cannot calculate with the heart 'what holds / the world together in its inmost folds', as Goethe put it. But we need a heart that has the urge to make such calculations and which can observe their mathematical beauty. What we feel in the heart also becomes conscious in the heart, as quantum physics does not stop in the brain.

Welcome to reality

The heartbeat of time was suspended. I was still standing, spices in hand, at the open-air stall of the organ dealer in Saigon. I was not bothered by the hundreds of flies buzzing around me. And even less by the billions of neutrinos that were pervading me. Neutrinos, by the way, come in three 'flavours': electron, muon and tau. This is what physicists call the different states which can oscillate into each other. Famous quantum physicist Wolfgang Pauli would not have suspected this when he postulated the existence of neutrinos in 1930. He was known for his joie de vivre but bemoaned the neutrinos: 'I have done a terrible thing. I have postulated a particle that cannot be detected.'[14] I felt for him, for was I not doing something similar? But I am sure of it, with my brain, my heart, my cardio-cognitive consciousness: the heart is an organ of consciousness.

'Mister, mister!' Someone was pressing a banknote into my hand. I had not noticed it slipping from my pocket when I paid for the spices. 'Thank you,' I said, 'thank you so much,' and returned completely to the reality of the market in Vietnam, a country which has ten words for 'heart'.

HEART ENCOUNTER

My heart had beaten countless times since I had sallied forth in search of the whole heart. The journey had changed me, and sometimes I wondered if I really wanted to keep going 'only' operating on hearts. I did not know how to reconcile my new findings with my old profession as a heart surgeon. It was my profession, after all, to operate. I could not very well issue my heart-deaf patients a prescription to use the HeartMath method to tune in to themselves more consciously. But it increasingly pained me when I sensed that a sick heart had been 'banged up' like a car by an inconsiderate lifestyle. To avoid the repair shop (operating theatre), the heart as the motor of our life needs a service every now and then. And ideally it should also be driven lovingly and attentively. As it is always beating and never rests, the service could be a service to ourselves: slowing down! Change into a lower gear and listen to our inner voice.

Sometimes when I look into the face of an exhausted heart during an operation, I wish that this motor had been treated more gently and its driver had better listened to its voice. But in the everyday business of a clinic, such therapeutic instructions from the mouth of a heart surgeon would have caused consternation – even though we can demonstrably not only calm our mind with meditation and breathing techniques, but also the heart. The effects on blood pressure, heart rate and brainwaves can be measured even in beginners and even on the first day. In a current scientific statement, the American Heart Association explicitly supports meditation, as it can lower the risk of cardiovascular disease. You can't conquer such diseases with technology alone. Even though every year many hundred billion dollars are spent on it, cardiovascular disease remains the main cause of illness and fatalities.[1]

If we – and the society in which we live – perceive our heart as a mere pump, rather than as our most important sensory organ, it becomes sick. And so does the brain. Depression is the most frequent reason worldwide for the impairment of human life, and in 2020 it was the second-most frequent cause of death.[2] Part of the reason for this is the fact that heart and brain are so often treated in isolation, while they belong together inextricably. What they would need is 'couples therapy' to uncover the disturbances of their reciprocal perception and communication.

I still valued my profession, but I now saw it as too one-sided when practised in a big clinic. I wanted time away from my green theatre scrubs, and, more than anything, I did not want to reduce a person to the small rectangle of flesh I saw between sheets of sterile green cloth. My life had altered considerably since I had started to listen to my heart more: my hobbies had changed and also my relationships with family members, friends and patients. But sometimes it took a lot of strength to hear the heart's voice amid the everyday proceedings of the clinic. When I returned after a weekend of meditation in a monastery or from a mindfulness seminar, from Monday morning I was expected to function and switch off my emotions in this pure-white high-tech environment – it did not fit together. It became clearer and clearer that as a heart surgeon I really wanted to treat the whole heart – especially the part I had found on my journey. I wanted to consider more than the excerpt under the green cloth – I wanted more time for my patients. I did not want to ask merely about their symptoms, but also to find out about their life circumstances, how they managed their everyday affairs, what worries plagued them: in short, what was weighing on their heart, because that plays a fundamental role in the heart's health. I am convinced that in the following years and decades we will find further proof that the heart also medically lives up to its symbolic importance. It feel-thinks or think-feels. I wanted not only to put a stethoscope to my patients' hearts, I wanted to listen to them actively. And my own heart, too – I wanted to get to know it a bit more.

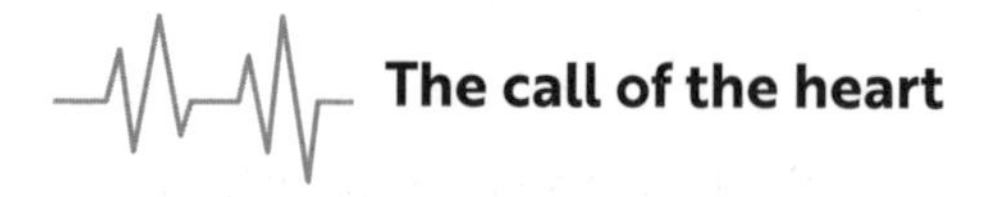

The call of the heart

For quite some time I had felt a longing in my heart to sail across the big sea. As a small boy I could not get enough of the magnificent sailboats on Lake Constance, which I visited on rare excursions with my parents on Sunday afternoons in the back seat of a Volkswagen Beetle. Without safety belt or child seat. But an exclusive hobby like sailing was unreachable for me back then. Instead, my mother gave me a globe, which I still have today. Back then, I went on an adventurous journey around the world with my friend Jimmie. We were free, we were rich, we spun the ball, stabbed it with our fingers and in our imagination drove, flew and sailed to foreign countries.

I had been a heart surgeon for years when I finally learned to sail, and I loved it as much as I had thought I would as a child. That was lucky, as some childhood dreams turn out to be rather unfulfilling in reality. But water was my element, and when the wind filled the sails, my heart would expand. A new longing grew. My family encouraged me to fulfil this dream: to sail across the Atlantic. After a long wait I was able to take my annual leave and saved-up leave from previous years in one go, two months in total. I had got to know my three co-sailors on the chartered yacht some weeks earlier. I got along well with all of them, but I liked Harald the skipper, a retired taxi driver, best. There was also an IT expert and an engineer on board. We set off in February from the small island of Saint Lucia in the Caribbean and only got to Antigua, where our motor broke down. Was it the motor? No. The surgeon we called, in oil-stained overalls, diagnosed a problem with the pump. The fuel injection pump had been irreparably damaged. We would need a transplant. So began the wait for the new pump. Daily we checked the UPS tracking to follow the route of the pump, which was coming from Europe. The crew's mood was strained: all but the retired skipper had taken time off for the cruise and the time pressure was enormously stressful. On the third day, I started a scuba-diving course. The underwater world fascinated me so much I no longer minded the wait – which the engineer and the IT expert didn't like as they did not share my good mood. After seven days, the IT guy announced that he could not stand it anymore and wanted to go home immediately. We tried to talk him out of it but soon realised it was pointless. Only now he confided in us that he suffered from panic attacks. He had believed that the stay on the boat would cure him. He had been wrong – he felt like shit. So there were just the three of us now, which the engineer did not like one bit. He complained excessively about the 'nasty, false, dishonest' IT guy who had misled us irresponsibly – he felt that it was his fault that our life dreams were falling through. I could not get the engineer to understand that panic attacks are an illness. He wanted to be right. Which can also turn into an illness. Or did he, too, merely need a reason to chicken out? The next morning he informed the skipper and me that he deemed the cruise too dangerous and was calling it quits. The two of us could set sail if we wanted to.

At this point, our boat was already fully supplied with provisions. Apples and potatoes were becoming old and rotten; the banana plant we had tied to the mast was dropping its fruit. And now? I wanted this cruise at any cost. The skipper thought it was possible for the two of us to do it, because I was an experienced sailor – but he was still a little uneasy about it, so we put up an ad: 'Sail to Europe for free'.

In Antigua, where the super-rich berth their superyachts, we were more in the category of social housing, or better: boating. But there were also some young people working as crew members on the yachts, and many travellers. One of them was Solomon from Israel, in his early thirties. And eventually John from New York joined us. He had become stranded in the Caribbean at some stage, broke and without any prospects. At least with us he would not have to worry about his meals for the next few weeks. He had a little bit of sailing experience. Solomon had never been on a boat before, as he freely admitted. 'Only once, with my father, on a pedal boat.'

Once the fuel injection pump had been replaced, we set sail, or at least we wanted to. But our anchor chain was blocked by the chains of the superyachts, and it took half a day for the divers to free us. Gradually I began to ask myself why we had been 'told' to stay in the harbour. Was there a deeper meaning to this? That's what you start to think when you have been delayed a number of times. And, after all, our voyage was by no means without danger. The sea does not forgive any mistakes – all who know it say that – and we faced a transatlantic passage of over 2000 nautical miles in a fifteen-metre boat against the prevailing wind direction. The Atlantic is a wild, limitless expanse of nature. For a long stretch of time one is far beyond the life-saving reach of civilisation – a veritable adventure, more dangerous than my travels around the globe with Jimmie. For me, though, the adventure unfolded very differently from how I had hoped. As soon as we had reached the open sea, I became seasick. Green and pale in the face, I was on the ropes. I was miserable and no help to the crew. To make it worse, I remembered how impatiently and insensitively I had reacted when my children or my wife went green around the gills during our holiday cruises in the Mediterranean. After three days I finally felt better, and the sea had also calmed down. Instead, John lost his mind. He was always walking around with a metal cup in his hand. At some point I saw him pour Campari into the cup. At eight in the morning.

'What are you drinking?' I asked him.

'Rosehip tea,' he replied.

'Smells different to me.'

'Yes, okay. Red wine.'

Some hours later he went completely ballistic. He bellowed and staggered across the deck, nearly fell overboard, lashed out and insulted us.

'He is completely drunk,' Harald diagnosed. He inspected the alcohol supplies and noticed that John had blasted through two-thirds of our supplies in less than a week. We had an alcoholic on board. And now? Harald and Solomon were looking at me. It was obvious to them that this was a case for the doc. So I held a patient consultation with John. He freely admitted to being an alcoholic.

'Anything else?' I asked, expecting the worst.

'No, just boozing.'

I spoke with him for a while and at some stage during the conversation realised that my compassion was not just a pretence to appease him. I was with John with my whole heart and tried to find a solution for all of us. John was not the least bit aggressive, but instead contrite – which was probably because I honestly cared about him and was trying to help. He accepted my plan straightaway. He was to be allowed to keep drinking – but only when everyone was having alcohol, meaning with meals. So he wouldn't be tempted, I would store the rest of the supplies under my bunk. 'You are a sailor, John: you know what this means. A sailor's bunk is sacred. You won't touch it.'

'Aye, aye, sir,' he said with a wry grin and put his right hand on his heart. A person's word is worth something among sailors, as they are completely at one another's mercy. It certainly was not easy for him, but he managed. Our mainsail didn't: it frayed and had to be stitched up. A job for a specialist – me. John assisted me companionably.

Two days later Harald, usually rather sparing with his words, paid me a pleasing compliment. 'I'm really glad to have you on board, Doc. It would not run as smoothly if we didn't have a ship's doctor among us. And the way you stitched up the sail – haute couture!'

Well. Compared with my fine sutures in the operating theatre, this had been rather quick and dirty. Harald was an experienced skipper and had a good knowledge of human nature. But calling me a ship's doctor? That amused me a little, because I had wanted to be on a ship to get away from it all. But the words would not leave my mind ... ship's doctor. There was something about it, particularly since I loved the sea so much. And I also wanted to improve my scuba diving.

Waves

I sat alone on deck at night, the immense starry sky above me. Again and again I looked up to find our lodestar. The one that pointed us in the right direction, among the 6000 stars we can see with the naked eye. There are 100 billion galaxies, and 100 billion is also the average number of human brain cells. Is this a mere coincidence, or is the brain also a universe with infinite possibilities and spaces? At night in the vastness of the ocean, my thoughts became freer, my questions bigger. How small I was, and how vulnerable. Endless water. Endless depth, endless expanse, endless waves, endless blue, endless sky. I felt snug in this infinity, in the arms of the wind. We depended on it; it was the only force that could bring us home. The diesel tanks of sailing yachts this size are small and designed for short distances, not to cross an ocean of many thousand nautical miles. By and by I could hear from which direction the wind was blowing and if the sails were correctly positioned. I could hear it because Earth, like the heart, is bedded on a pillow of air. Sounds are air waves without which I would not be able to perceive the swooshing of the huge Atlantic waves crashing into the boat – or the heartbeat of my patients.

Without compass or helmsperson we would be lost, just as we would without brain or heart. The heart trembles within us like a compass needle, and if we pay attention to it, it will pilot us safely through life, in conjunction with the wind. You can get to this wind on many paths, with the waves, with the stars. The heart grants some degree of freedom. But you cannot sail against the wind. It would be foolish to try. You would arrive nowhere. Like Odysseus, a man who was looking his whole lifetime for a way home.

When I surfaced from such thoughts, I meticulously checked our course, which would take us to the other side of the Atlantic. Sometimes a wave threw salty drops of water into my face. I licked them off my lips and swallowed them. The salt content of the oceans is nearly identical to that of our blood. We come from the sea, and I felt in resonance with this ocean of waves. I. Tiny like a water drop. Where should I anchor? Who am I? A drop of water. What is my task in life?

Consciousness, for me, is a condensed drop of water. There are certain physical conditions somewhere and suddenly a drop of water is hanging from a leaf or running down a windowpane. Air humidity can't usually be seen. Only when condensation occurs do we see drops of water. In the same way, our life condensates from cosmic elements in an act of biological creation. We are visible drops in the ocean of being. I am a drop of water that moves, and inside me pulsates a heart. Every water drop will eventually find its way back to the sea. And at some point, a new small heart began to beat inside me here on the Atlantic . . . the heart of a ship's doctor.

When we reached the Azores my trip was over, because my leave was used up. I would have loved to simply sail on. But that's not me. My colleagues were waiting for me, and I would not let them down. Besides, I had arrived. Maybe not in southern Spain, but at a much more important destination:

at my self. I knew now where my heart was calling me. And that my time in the clinic was finished. Although I would return for now, I would not continue as a heart surgeon forever. The time had come to work for the whole heart – in the words of a character from Franz Werfel's novel *The Forty Days of Musa Dagh*: 'There are two hearts. There is the heart of flesh, and the secret, heavenly heart which surrounds the other, just as its scent surrounds the rose. This secret heart unites us with God, and other men.'[3]

And, besides, I wanted to find out what I needed to do to sign up as a ship's doctor. Because in that way I could connect everything. The sea and the sky, the heart and my patients, and myself.

In new waters

It took a while until I found the opportunity to sail in new waters. Everything was different from the moment I knew that I would change professionally. From then on, I no longer worried what others might think of me; my decision had been made with heart and mind, carefully thought through.

In my operating team, I placed even more emphasis on respect towards patients, even though they could not hear or see us. I thought of myself as the guardian of this room; the devout silence in the theatre had always reminded me of a church, and now I took responsibility not only for the vibration of the medical saw but also for the vibe in the room. I noticed that my attitude was rubbing off on my team. Our working together had never been as wonderful as in my last months. I told no one of my decision. I was still in a phase of finding myself and had a lot to consider. One day I read about Markus Studer, a heart surgeon from Switzerland who, in his mid-fifties and at the peak of his career, fulfilled a childhood dream and became a trucker. In an interview with the *ÄrzteZeitung*, he said: 'I did not want to operate for too long – that's not good, neither for the doctor nor for the patients. Especially in surgery doctors should put the scalpel down in their mid or late fifties.'[4] I, too, think that is a good age to follow the GPS of one's forgotten heart.

The additional training to become a ship's doctor was not a problem. But a second heart was beating in my chest, and it wanted to be heard as well. For a while, I had been carrying inside me the plan to develop a medicine for the whole heart. I would only be able to implement that if I had my own practice for holistic and operative heart medicine. I wanted to continue to be a heart surgeon, but play on heart medicine's whole keyboard. First and foremost, I wanted to take the time to find the true causes of my patients' heart complaints. Those you don't see at first glance. I was certain that some invasive procedures would then no longer be necessary. The price I would have to pay for this change was that, as an independent heart surgeon, I would no longer be allowed to conduct certain operations, for example implanting an artificial heart, my special field. Thus, these last months in the clinic were also a goodbye, in some way. I began to relish my time in the theatre, similar to my enjoyment as a young assistant when everything had still been very new. And it was indeed very new for me once more – as everything we perceive with an open heart is fresh and unique.

FAREWELL TO THE ARTIFICIAL HEART

The heart that lay before me was looking heavenwards. It was presumably the last one I would connect to an artificial pump. With several surgical drapes I had put the heart in the right position, so its apex was pointing up. It was connected to a heart-lung machine, but still beating. In the area of the apex I had stitched a metal ring, through the centre of which I was now punching a hole slightly larger than two centimetres into the left ventricle.

Ten days earlier, the patient had been admitted to the clinic with a severe heart attack, and cardiologists had implanted several small metal tubes – stents – in the blocked arteries of his heart. Twice he had suffered from ventricular fibrillation, twice he had been resuscitated. We succeeded in saving his life, but his heart remained weak. Too weak to continuously manage the blood supply to his organs. We call this state heart shock, or cardiogenic shock, and it is life-threatening. The body needs to be supplied with energy from outside. An established method of doing so is to first connect the patient to ECMO (extracorporeal membrane oxygenation), an emergency pump that supports the heart and oxygenates the blood. In the first operation, I had joined up two tubes nearly the diameter of a garden hose with the blood vessel system near the heart and had then connected them to an external pump system. Twice daily a team of heart surgeons, anaesthetists, cardiologists and nursing staff had gathered at the patient's bed and discussed the state of Mr Rubella and his heart. We had tried several times to reduce the ECMO support. Sometimes hearts recover after a few days and ECMO can be removed. Not so with this patient. The more the pump flow was reduced, the weaker his heart was beating. Everyone involved was able to witness this live via echocardiography, the ultrasound of the heart. Doctors would call the weaning unsuccessful. With the pump, the patient's condition was 'stable'. He was on dialysis because of kidney failure, but his other organs were doing their work relatively well, and the brain did not appear to be damaged either.

There were now two possibilities: to let the patient die or to implant an artificial heart pump made of metal, a VAD (ventricular assist device). Both options were justifiable; the question was what Mr Rubella would have wanted. I spoke with his wife and children. In considering implanting a permanent VAD, one needs to be aware of the considerably heightened risk during the operation with such a critically ill patient. However, if it was successful, his life would be completely different. A cable, the so-called driveline, would protrude from his left side under the ribs, and he would spend the rest of his life with batteries on his belt. I once had a colleague who said that such patients and their surgeons were 'married until death do them part'. That is ribald heart surgeon humour, but there is a grain of truth in it: because if everything goes well and the patient returns home, the clinic must be available to him day and night in case of emergency. My questions to this patient's wife and children were: would your husband and father have wanted a life with an artificial heart implant, and would he be able to manage it? Would he accept the technology and be able to change the batteries every day? Would he take the blood-thinning medication regularly, would he be able to live with a cable protruding from his side, and who would look after the resulting wound? Experience suggests that patients will not manage this by themselves; they need helping hands and lots – really, lots – of loving support. However, if they have this they can lead a life that is nearly normal. Some even still go on cruises. Mr Rubella was currently embarking on his final journey, his life balancing on a knife's edge. But his wife and children believed he wanted to live at all costs, and they wished to walk the difficult road ahead with him.

The medical team met again, and we decided to risk it. The timing was favourable, which is a decisive factor in such an undertaking. One must not wait too long. When the process of dying has begun, it is too late. No charts will tell you exactly when that is, but there are signs – and experienced doctors know the many faces of approaching death. And of course the VAD should not be put in too soon, while there are other options. This decision can be a huge challenge. After everything was settled, the operation was prepared for the next day.

Stopping the heart is not always necessary with these procedures, and now I was looking into the patient's beating heart through the punched-out hole. This spot has to be 'perfect' because the suction tube needs to lie free and central in the heart to enable optimal flow. Whether that has really worked will only become clear towards the end of the operation, but at the moment it was looking good. Inside the heart I found nothing that could have disturbed the blood flow through the metallic suction tube. The actual transport of the blood is done by an impeller in a small shiny chrome box that is round and slightly smaller than a palm. From there, the blood is passed on to the aorta with a flexible prosthetic tube. The professional term is left ventricular assist device (LVAD), a supporting pump for the left ventricle, which is mounted onto the heart. For decades there have been attempts to build artificial hearts that would replace the whole heart – apparatuses with valves and ventricles which produce a pulse wave, like the real heart. But the natural technology of the latter is unsurpassed, and the engineered 'total artificial heart' often leads to severe complications. So engineers and heart surgeons thought again and constructed something that is very simple but functions better. The old heart remains where it is, the patients survive for longer and there are fewer complications. Perhaps the advantages compared to the 'total artificial heart' also come from the fact that the connections with the body and self-perception can remain intact.[1] Even with transplanted hearts it is possible that severed nerve tracts connect again. And the more they do that, the better the transplanted heart will work.[2]

I lowered the suction tube into the heart, checked again if the box was sitting on the metal ring as planned and fixed it with a single small clamp screw and a torque spanner. Thorough ventilation followed, as there must not be any air where there should be blood. This routine took at least ten minutes. In the meantime I recalled, as I had many times, the moped (a NSU Quickly) I had owned as a school student and whose motor had so often needed repair. It had always been important to thoroughly vent the carburettor, and once after I had overwound a screw I had bought a torque spanner from my pocket money, not imagining that this tool would one day serve me well during heart operations.

There are some patients who, after an operation, have enormous psychological problems living with a machine in their chest. Some even cut through the cable to the batteries, as they cannot pull the plug for safety reasons. I remember an elderly lady especially well. She considered the 'steel box' in her chest a foreign body and wished for nothing more than to get rid of it again. For her four-year-old granddaughter, however, the box was the 'darling'. It was due to the box that her grandma was alive. And since she was always asking grandma about her 'darling' and if it was still humming well, the iron heart in her chest became grandma's darling too. The sonorous hum, by the way, is a calming sign for patients. If it starts to rumble and creak, that indicates a pump thrombosis – a grave complication. One thing I had learned even back then with my Quickly: before the motor stopped and I had to push the vehicle for kilometres, there was always a signal indicating a problem – and if I paid enough attention I could hear it.

Fortunately the turbine in Mr Rubella's heart had started to run smoothly in the meantime, and all air had been removed from the heart. His heart was filled, the length of the prosthetic tube measured and implanted into the aorta. What now follows is a subtle process in which the LVAD flow is increased gradually and that of the heart-lung machine simultaneously reduced. The right amount of the appropriate medication is necessary for this, as well as lots of instinct and patience. One has to give the heart time to get used to the new situation.

The body adjusted to life without a pulse without any problems. Even though the heart is still beating, the blood is now transported to the body by the turbine, with its steady flow. Supported by the heart-love-and-life hormone adrenaline as well as numerous other drugs, the patient was brought to the intensive care department. He suffered from heavy loss of blood for days, and twice his chest had to be opened again to look for causes of the bleedings. Not because the operation had been sloppy, but because this patient's clotting was 'going down the drain'. The lost blood is caught in special containers and infused back into the patient. It is a battle of material against death in which modern medicine comes out all guns blazing. The worrying and hoping lasted for two weeks, and then Mr Rubella reached calm waters, so we could reduce the anaesthetic and let him wake up. His brain had not suffered any damage. He recognised us. Those are amazing moments for patients, family members and medical team. After more than a month the patient was able to go for the first small walks into his new life.

Setting sail

And for me, it was time to set sail. To become a ship's doctor, I had to go back to school. Together with apprentice ship-builders, as well as fully qualified engineers and doctors, I completed several safety training courses. I had learned to scuba-dive in the Caribbean, but now I added the training for diving doctors. My adventurous heart was always glad when we were drifting in wintery water in survival suits or experiencing the rapture of the deep with our own bodies in the pressure chamber.

From the moment I had begun to follow my heart, many wonderful things had happened. But in some dark hours I doubted my decision. Was it really right to change waters? Every now and then I was pestered by existential fears. How would it all unfold once I left the safe harbour of the clinic? In such moments I had to tell my heart to pull itself together, or else my heart told *me*, in a very healthy way. Again, I experienced that it is the most beautiful thing to be in resonance with my environment and the people around me, trusting them. This trust stayed with me during my training to become a ship's doctor, and when I became self-employed with my own surgeon's practice – because I didn't know if I would be needed as a ship's doctor, nor if any patients would find me. Both found me, ship and patients. And further wonderful things followed. During a medical seminar I told a colleague about the concerns of my heart. Shortly afterwards I received an invitation to give a talk on the topic of heart and brain. This would be my first public appearance on the topic of my heart.

I commissioned a website for my practice. The designer told a friend who was a journalist about my mission of the 'whole heart'. Before long I was interviewed for a medical publication in *Die Zeit*. The article created a stir, I received a lot of mail, a magazine asked me for an essay. And so we had come full circle. Many years ago, a patient had given me the book *The Wise Heart* by Jack Kornfield and inspired changes. And now I was side by side, so to speak, with this teacher of wisdom – my essay and an interview with Jack Kornfield in the magazine *moment by moment*. What an honour! This drew the interest of several publishing houses, who asked me if I would be interested in writing a book. And now you are reading the result. What will you do with it? Which circles will close for you, which ones will you open – and where will the voice of your heart carry you?

Mine was calling me to sea, to the place of my longing.

The hungry heart

Those who go to sea need an anchor. That is what my practice is for me. Here I have the time and energy to care for the hearts of those seeking my help, who at times drift lost in the sea of high-tech heart medicine.

One day I received an email from Astrid, a young woman from Austria. Her heart was no longer pumping effectively, and she might need a transplant soon. I learned from scans from several university clinics that she had suffered repeatedly from myocarditis. Histological examinations of her heart had shown scars. The right side of the heart was especially affected, the ventricle was widened and the tricuspid valve between ventricle and atrium did not close properly. All of the examinations that were possible on the heart had been conducted. The patient was taking numerous drugs, but her condition was getting worse and worse. She had trouble breathing, and her legs were swelling. So after her last stay at a clinic they had suggested she visit a transplant centre. The conversation with the chief physician there had been very matter-of-fact and had only lasted a few minutes. After that she had cried for days and felt completely abandoned ... and then she asked if I could help her.

A week later she came to my practice. We spoke for a long time, and she told me things you would not find in a typical doctor's letter. Since her late childhood and far into her youth, she had suffered from bulimia. Her mother had died when giving birth to Astrid, who had been brought up by an aunt, who had not been able to give her the love and care she needed. In her family it had been common to express affection and warm-heartedness through cooking. At some point the girl began to throw up, and after a few years her heart had become inflamed. Ever since then her heart would often race after meals, even though she had overcome the bulimia.

I examined her heart via ultrasound, and we watched the images together. I asked her what she was feeling. 'I hate my heart,' she replied. 'I hate it because it isn't functioning, but I am.' I asked her to feel her heart. She refused. 'Feeling it is exactly what I *don't* want to do, I want it to leave me alone.' After I asked what her heart needed, she was silent for a long time. Then she started to cry.

My first check of her heart rate variability confirmed on a physiological level that Astrid and her heart were no longer a team. She stayed for several days, and I asked her to try to observe her heart rate during her walks along the Baltic Sea coast. Just perceive, not judge. She trusted me and tried it. On the next day, she said: 'I think I know now what my heart wants.'

'Yes?'

'It wants to be loved,' she sobbed.

'Yes,' I said.

Over the next few days we started the therapy, and it turned out that Astrid was able to support her heart's pump function with conscious, loving breathing. She called it: 'I embrace my heart.'

Within a year, Astrid came back twice to her 'heart time at the Baltic Sea', as she called it. She was much better now. She was able to do yoga again, did not have so much trouble breathing and even wanted to start going to work again. This positive development was confirmed by the follow-up MRI scans of her heart. Over time, the pump function even increased.

Psychoneuroimmunology also deals with patients like Astrid and verifies how much our inner experience can affect our immune system. I believe the key to Astrid's heart healing was her willingness to feel herself and her heart.

The biggest health issues of our time – depression, addictions of all kinds, and incurable pain conditions – can all be traced back to a disrupted ability to feel.[3] I am convinced that is true for many illnesses of the heart too, and that we should embrace our heart every now and then to get closer to ourselves, to feel ourselves and also our fellow human beings.

HOMO COR

In the history of evolution, the ability to feel oneself is seen as the beginning of our consciousness.[1] Evolving human beings had at one point in their history and the history of the world felt their own heartbeat and how 'something' inside was alive. A few million years have passed since then, the brain has grown and we call ourselves *Homo sapiens*, the knowing human. There is, however, a central question that has remained completely unanswered: how is subjective experience created within our bodies and where does consciousness come from? No existing theory can be verified metaphysically and empirically.[2]

Findings from quantum physics allow us to conclude that our borders do not end at the surface of our skin, but that we are connected with everything in existence. Many people are familiar with this feeling of being 'connected' – with nature, the universe, a beloved person or animal, perhaps even with a special purpose in life. The findings of multiverse theory, which unites physical astronomy and quantum mechanics, suggest that there are many universes with many different versions of physical laws.[3] Furthermore, according to the official theories of physics, we live in an eleven-dimensional entity whose directly perceivable dimensions are three-dimensional space and time. Hawking and Mlodinow rightly ask in their book *The Grand Design*: 'If they are present, why don't we notice these extra dimensions?' and answer this question straightaway: 'they are curved up into a space of very small size' so we do not notice them.[4] Is that really the case?

Life on Earth is 3.8 billion years old, and the first prehistoric person lived seven million years ago. I think it is very unlikely that after this long time of human evolution and life with nature there are dimensions that we cannot *perceive*. Nature has had lots of time to equip us with multiple possibilities of perception. I am therefore convinced that we *do* perceive the dimensions of M-theory. Love and compassion, truth, wisdom, strength, joy and thankfulness may be some of them. In short: the consciousness of the heart. Not every one of these qualities is spontaneously available to most of us, but if they become feel-able or 'feel true', if they unwind and unfold, they open a space, a dimension, their own universe. They are not mere 'feelings', but in their full extent are dimensions of humanity. And humans are not humans only because they are cognitively active and have a brilliant brain – but also because they have their heart in the right place. We have developed into *Homo sapiens* precisely because we feel the dimensions of the heart.

So we are not only knowing humans but also heart-centred humans. *Homo cor.*

Nature has made the creative, enterprising and feeling human being as complex as the universe itself. If even a single star were removed from the latter, we would not have the conditions on Earth which we need to live. With a human being, we must not remove the feeling heart, otherwise we would vegetate at an intellectual level. That would not be what I call being alive. Life only makes *sense* if we use our *sensory* organs to their full extent. Without the heart we would not have created magnificent works of art, would not have developed visions of humanity and not have made scientific discoveries. The following is my incomplete chain of evidence: Beethoven's symphonies, AC/DC, the cave paintings from the Stone Age, Picasso, Gandhi, Anne Frank, Einstein, my children, Yin and Yang, Marie Curie, Schopenhauer, Czech children's TV series *Pan Tau*, Freud and Porsche. What does your chain of evidence look like?

With the whole heart doesn't mean brainless

In the last few years there has been a lot of talk about *Homo deus*, the divine human, who gains god like abilities through technology.[5] But where has that led us? We can destroy ourselves at any time, can poison our Earth irreversibly. With our brilliant minds alone, we are not *Homo deus*: rather, the opposite. The divine in us can only unfold if we integrate the heart into the achievements of our mind. Critics may object that the heart's voice has never been measured by anyone. But the age of science and mechanisation – from the steam engine to the space station, the letterpress to Facebook – has only been going on for a few centuries, which is a very short time compared to millions of years of evolution. Thus, it is understandable that in this short time we have not yet developed measuring instruments for all dimensions that we humans can perceive. It is no surprise at all that we can't yet measure our body's complete sensor system for those dimensions. Maybe that system is simply *immeasurable*, like the extent of the universe. But we should be wary of claiming that such sensors do not exist and that love, wisdom, compassion and the many other essential perceivable qualities of the heart are 'only' feelings. If one views them with all their possibilities and facets, in their meanings and how they influence our lives, they have the extent of dimensions we can experience with our senses and which become conscious in the heart. Theoretical physics has now calculated the existence of such dimensions. Does that mean we can express in numbers that which we perceive? Basically it can be explained quite simply using an example: we can immediately perceive three-dimensional space with our senses, or we can depict this space mathematically. Both are correct, but they are not the same. There is a tremendous difference between measuring the water temperature in a bathtub with a thermometer, and lying in a warm tub and feeling it with all our senses. They are different representations of the same truth.

The heart is not the storage location for the things we can feel with it – for hearts can be transplanted, and that does not influence (or only to a very small degree) our love for certain people. If a part of your TV is changed, you do not usually receive different stations afterwards – but you can see an image again. The heart is a biological processor for love, not its container. In the same way, the brain is not the container for parts of consciousness such as intelligence or creativity, but rather an essential organic constituent in the process of the emergence of these dimensions. The topography of consciousness is alive in every one of our body cells and also outside of our body. We are permanently surrounded by consciousness. No one knows if we will ever decipher its emergence. But I am certain that the code is to be found in the scientific understanding of the body as a whole, not only in the brain. Body-centred therapies and spiritual movements have known for a long time that one can gain access to consciousness through the body, and they work with that. Humanity as an act of loving care and empathy is not measurable with an array of technical instruments. In this regard, life is ahead of technology by a few million years of evolutionary development. It takes a heart and a brain to understand that.

What is currently in vogue, however, is the brain alone. Those who still bet on the heart, in the face of this brain-dominated idea of humans, appear to make fools of themselves. All power comes from the brain; it is said to be the control room, our supercomputer. We do not understand how it works, but have an age-old wish to improve it. And so we research it intensively to determine how we can make ourselves more efficient with drugs, virtual realities and brain electrodes. There is a whole industry of so-called neuro-gadgets, tools and artificial body parts, which are designed to optimise our thoughts, feelings and memory. Their inventors think we are far from utilising the brain's full potential. Car manufacturer Tesla, too, is tinkering with a digital Nuremberg Funnel of the new age. With their firm Neuralink, they want to lay a digital data highway into the brain in order to link it with artificial intelligence. I deem it impudent to want to gain, with surprising transparency, control of our thinking in order to degrade us into manipulated bio-robots. Would it not be much wiser to link our brain with true intelligence, which comes from the heart as well? Biological problems need biological solutions, and you cannot 'hack' a heart. Artificial intelligence is heartless and therefore, in my view, can never be intelligent. It cannot replace heart intelligence – with which miraculously we are already equipped. We do not have to undergo an operation, do not have to swallow brain-performance pills (so called neuro enhancers) and do not have to put on digital hats for transcranial magnetic stimulation. In any case, these technologies are positively primitive compared with the most intelligent organ of the universe, our brain. Humankind did not create the computer in its image. Because it can't, as it would be far too complicated. That is why the interface where nerves meet plates, consciousness meets software, the sound of thoughts meets transistors remains a problem which is completely unsolved. As opposed to the connections from heart to brain, which have already been laid in multiple ways by Mother Nature, and the sensors are online. Shall we receive and feel? In view of the clear physiological signals of the heart to the brain and the measurable responses of the latter, it would be a deeply rational decision and a sign of true intelligence to use this ability to feel that nature has equipped us with. It would be a strategically wise decision to use the potential fully, instead of relying on questionable brain optimisers. To listen to the heart is not a 'feeling' in the sense of a nebulous, emotional vagueness. On the contrary: it brings clarity.

The no-heart syndrome

Unfortunately, I am observing an epidemic that is rampant worldwide: the no-heart syndrome, or heartlessness. A flexible heart, compassion and humanity have been lost to those affected. They suffer from a rigid fixation on the brain and from a deficient ability to feel. Self-love and emotional regulation are lacking; manipulation and lies dominate. Our heart is a sensitive instrument with which we can distinguish true from false. Listening to the heart does *not* mean falling dumbly and with blind trust into every trap or being led by corny emotions. The heart is not the department of love, peace and harmony in our chest. Decision-making led by the heart is clear, life-affirming and never populist. To implement the heart's messages can therefore require a lot of courage. Without his warrior heart, David would never have beaten the far superior Goliath. We can only win a fight when our heart mobilises every single cell in our body. The heart loves adventures that the brain alone would be frightened of. But together they are an unbeatable team. The mind, on its own, thinks and thinks but does not know the truth. It has no access to the secret of life, whose code will only reveal itself when heart and brain come together – when the two pieces of a treasure map fit together. Then the gate to knowledge will open.

Many executives have realised that they cannot do without the resources of the heart in order to be successful, work sustainably and have satisfied and motivated staff. Only when intelligence is paired with the ability to feel will knowledge become wisdom and the high-performer become the heart-performer. The latter takes responsibility for creation and is not merely motivated by greed. We should take responsibility in the heart for what we do and also for what we don't do.

How do you feel with your heart?

Most of us feel it: without the heart we are not complete. But how exactly does it work, feeling with the heart? How do you act from the heart? It is really quite simple. It is innate and part of our true nature. Should you be curious and want to go deeper, I recommend you ask yourself two questions.

What does your heart really need? Possibly you haven't asked yourself this question in a long time. Allow all feelings that well up. Whatever you feel, it is right, as it comes from your heart.

But be alert: thinking is not feeling! I have learned after many detours that perceiving is not the mental representation of feeling, but immediate experience.

You can explain to someone in words what it is like to be in love, to enjoy this heavenly feeling. But if they have not experienced it themselves, it remains a theoretical process, which can of course be so intense that they think: *oh, now I know what being in love means*. But that is a false conclusion, as the sweet amalgamating experience that hits when one is together with the beloved takes place in another dimension. It unfolds not only in the brain but in the whole body.

It is not only pleasant topics that can surface in the heart. Maybe there is something to regret, and we wish we had acted differently in certain situations. To condemn oneself in hindsight would mean to misunderstand the principles of the heart. Compassion for oneself and one's shortcomings is more appropriate. Self-compassion is the basis of compassion for others. It is the humanitarian aid of the heart. We are all imperfect, and it is wonderful to learn to accept oneself instead of looking for permanent optimisation. The perception of the heart has the ability to heal and paves the way to more clarity and self-love. And then a change in lifestyle – more exercise, less food, whatever it may be – will simply come in time, because we know we are worth it.

The second question is: what does my heart have to give today? A smile, a nice word, a gentle touch, a well-disposed wish? Whatever it may be, do it! A minute of such heart 'jogging' every day is a start. And it is good for the pump too.

Heart time

Nicolaus Copernicus discovered a long time ago that the Sun, rather than the Earth, is the centre of our solar system. This marked the end of the geocentric worldview and had countless repercussions for philosophy and natural sciences. The modern age was dawning. For me, the heart is the sun in the planetary system of our body. It rises in the morning of our life, when a few drops of love hormone encounter undifferentiated stem cells. It gives us heart warmth, also when we are alone, without which we could not live. I believe the moment has come for a Copernican turn of the heart and a new time. Heart time. It marks the end of the dominant neurocentralism. A time in which people realise that consciousness is a symphony which is orchestrated in the whole body and in the universe. In it, our brain and our heart circle each other inextricably in their own cosmic orbits.

One question did not leave me alone for a long time. All organs need a break at some point. Even the brain wants to sleep every now and then. Besides, it likes to work the most when it feels like it and sees the possibility of a reward. That's how it is for me, anyway, and numerous studies confirm that as well.[6] But the heart never has a break. How is that possible? How is the heart capable of this unimaginable achievement of pumping for a lifetime? For this book, I have advanced into the depths of scientific literature, but did not find an answer to this question. Whomever I asked – and there were fantastic medical professionals among them – no one was able to give me a conclusive answer. I have long thought and meditated about it. The more I relaxed, the closer I came to the riddle's solution: the heart always relaxes! After every single heartbeat, the heart relaxes. The time it sets aside for relaxation (when at peace) is about twice as long as the time for a contraction. If our heart has to beat faster, its breaks become shorter. But in any case it takes a break – until the next heartbeat arrives. So the heart not only contracts three billion times in seventy years, but relaxes as often. If you add it up, it relaxes far longer than it works. And from that, I believe, we could learn something. Peace is not only positive and work not only negative. One does not exist without the other – they are inseparably connected as the Yin and the Yang, life's beginning and its end. I believe this to be one of the really big secrets. Everything that is, from the Big Bang to the blinking of your eyes which may be tired now, follows an eternal rhythm of exertion and relaxation, the principle of the heart and of life.

I want to farewell you with an age-old custom of heart contact, a quote from the novel I mentioned earlier: *The Forty Days of Musa Dagh* by Franz Werfel. And as a doctor I want to add: this kind of greeting is more hygienic than handshaking.

'This old man's little hands waved in ceremonious welcome; they touched his heart, his mouth, his forehead. Gabriel was equally ceremonious. No impatience would have seemed to tighten his nerves. The Agha came nearer and stretched forth his right hand towards his visitor's heart, so that his fingertips just rested on Gabriel's chest. This was the "heart-felt contact", the closest form of personal sympathy."[7]

Ba-boom, ba-boom, ba-boom
Heart time is the present. Now.

Ahoy!

ACKNOWLEDGEMENTS

I thank Regina König and Hellwig Schinko for their loving company during the search for my heart. Olivia for reading the manuscript and her heart for the word. Josef for his heart music, which always sustained me. Heiko Köppke for his support in difficult times. Doreen Kuempel for her ideas on the electromagnetism of the heart on steel ships. Dr Markus Peters, who gave me an understanding of the spiritual medicine of the heart. Dr Detlef Reineck for his views on sensory perception and the meaning of life. Bernd Engler for his observations on the heart as a kinetic work of art which oscillates between feeling, being felt and feeling with someone. Finally Klaus Marsiske, because his images made tangible that which has no dimensions. Shirley Michaela Seul for her friendship and devotion to this book.

ENDNOTES

Revealing the Heart

1 N.H. Bishopric, 'Evolution of the heart from bacteria to man', *Annals of the New York Academy of Sciences*, vol. 1047, June 2005, pp. 13–29.

2 E. Alexander, 'Interview: On the mensuration of eternity', www.randomhouse.de/Eben-Alexander-im-Interview-zu-Vermessung-der-Ewigkeit/Interview-mit-Eben-Alexander/aid66622_13041.rhd, accessed 16 March 2020; B. Merker, 'Consciousness without a cerebral cortex: A challenge for neuroscience and medicine', *Behavioral and Brain Sciences*, vol. 30, no. 1, February 2007, pp. 63–81; discussion pp. 81–134.

3 H. Luczak, 'Neurologie – wie der Bauch den Kopf bestimmt' ['Neurology: How the gut rules the head'], www.geo.de/wissen/13364-rtkl-neurologie-wie-der-bauch-den-kopf-bestimmt, accessed 15 March 2020.

4 E. Craige, 'Should auscultation be rehabilitated?', *The New England Journal of Medicine*, vol. 318, no. 24, 16 June 1988, pp. 1611–13

5 M.K. Heinemann, 'Heart murmurs', *The Journal of Thoracic and Cardiovascular Surgery*, vol. 66, no. 5, August 2018, p. 359.

The Six-Cylinder Bio-Turbine

1 G. Buckberg et al., 'Structure and function relationships of the helical ventricular myocardial band', *The Journal of Thoracic and Cardiovascular Surgery*, vol. 136, no. 3, September 2008, pp. 578–89.

2 C. Song et al., 'Cardiac scan: A non-contact and continuous heart-based user authentication system', https://sctracy.github.io/chensong.github.io/pdf/mobicom17.pdf, accessed 16 March 2020.

3 C. Goddemeier, 'William Harvey (1587–1657). Die Entdeckung des Blutkreislaufs' ['William Harvey (1587–1657): The discovery of blood circulation'], *Deutsches Ärzteblatt*, vol. 104, no. 20, 18 May 2007.

4 R. McCraty et al., 'The coherent heart: Heart-brain interactions, psychophysiological coherence, and the emergence of system-wide order', *Integral Review*, vol. 5, no. 2, 2009.

Heart on the Table

1 L Meng et al., 'Cardiac output and cerebral blood flow: The integrated regulation of brain perfusion in adult humans', *Anesthesiology*, vol. 123, no. 5, November 2015, pp. 1198–208.

Chimeras of the Heart

1 L.K. Jones et al., 'Ethological observations of social behavior in the operating room', *Proceedings of the National Academy of Sciences of the United States of America*, vol. 115, no. 29, 17 July 2018, pp. 7575–80.

2 M. Garcia et al., 'Cardiovascular disease in women: Clinical perspectives', *Circulation Research*, vol. 118, no. 8, 15 April 2016, pp. 1273–93.

The Colourful Neuroshow

1 G. Santoro et al., 'The anatomic location of the soul from the heart, through the brain, to the whole body, and beyond: A journey through Western history, science, and philosophy', *Neurosurgery*, vol. 65, no. 4, October 2009, pp. 633–43; discussion p. 643.

2 J.J. Loizzo, 'The subtle body: An interoceptive map of central nervous system function and meditative mind-brain-body

integration', *Annals of the New York Academy of Sciences*, vol. 1373, no. 1, June 2016, pp. 78–95.

3 T. Fuchs, 'Kopf oder Körper? Dem ich auf der Spur' ['Head or body? Tracing the "I"'], *Draußen & Drinnen*, no. 5, 2014: https://heiup.uni-heidelberg.de/journals/index.php/rupertocarola/article/view/17268/11083, accessed 16 March 2020.

4 T. Fuchs, 'Verkörperte Emotionen. Wie Gefühl und Leib zusammenhängen' ['Embodied emotions: How feeling and body are connected'], *Psychologische Medizin*, vol. 25, no. 1, 2014.

5 P. Hummel, 'Hirnforschung im "Human Brain Project", Dicke Schädel, falsche Versprechen' ['Brain research in the Human Brain Project: Thick skulls, false promises']. *Süddeutsche Zeitung*, 1 May 2015.

6 U. Schnabel, 'Hirnforschung: Die große Neuro-Show' ['Brain research: The big neuroshow'], *Die Ziet*, 20 February 2014.

7 Arthur Eddington, *Wikiquote*, en.wikiquote.org/wiki/Arthur_Eddington, accessed 16 March 2020.

8 Fuchs, 'Kopf oder Körper?'.

9 A.K. Fetterman & M.D. Robinson, 'Do you use your head or follow your heart? Self-location predicts personality, emotion, decision making, and performance', *Journal of Personality and Social Psychology*, August 2013, vol. 105, no. 2, pp. 316–34.

10 Santoro et al., 'The anatomic location of the soul'.

11 B.N. Justin et al., 'Heart disease as a risk factor for dementia', *Clinical Epidemiology*, vol. 5, 2013, pp. 135–45.

12 P. Taggart et al., 'Heart-brain interactions in cardiac arrhythmia', *Heart*, vol. 97, no. 9, May 2011, pp. 698–708; P. Taggart et al., 'Significance of neuro-cardiac control mechanisms governed by higher regions of the brain', *Autonomic Neuroscience*, no. 199, August 2016, pp. 54–65.

13 Taggart et al., 'Significance of neuro-cardiac control mechanisms'; M.A. Samuels, 'The brain-heart connection', *Circulation*, vol. 116, 2007, pp. 77–84; K. Shivkumar et al., 'Clinical neurocardiology defining the value of neuroscience-based cardiovascular therapeutics', *Journal of Physiology*, vol. 594, no. 14, 15 July 2016, pp. 3911–54.

14 Santoro et al., 'The anatomic location of the soul'.

15 M.E. Ceylan et al., 'The soul, as an uninhibited mental activity, is reduced into consciousness by rules of quantum physics', *Integrative Psychological and Behavioural Science*, vol. 51, no. 4, December 2017, pp. 582–97.

16 F. Shaffer et al., 'A healthy heart is not a metronome: An integrative review of the heart's anatomy and heart rate variability', *Frontiers in Psychology*, vol. 30, no. 5, September 2014; R. McCraty & M.A. Zayas, 'Cardiac coherence, self-regulation, autonomic stability, and psychosocial well-being', *Frontiers in Psychology*, vol. 29, no. 5, September 2014, p. 1090; J.L. Ardell et al., 'Translational neurocardiology: Preclinical and cardioneural integrative aspects', *Journal of Physiology*, vol. 594, no. 14, 15 July 2016, pp. 3877–909.

17 M. Parzuchowski et al., 'From the heart: Hand over heart as an embodiment of honesty', *Cognitive Processing*, vol. 15, no. 3, August 2014, pp. 237–44.

18 J.R. Doty, *Into the Magic Shop: A neurosurgeon's quest to discover the mysteries of the brain and the secrets of the heart*, Avery, New York, 2016.

19 D.E. Ingber et al., 'Tensegrity, cellular biophysics, and the mechanics of living systems', *Reports on Progress in Physics*, vol. 77, no. 4, April 2014; S. Chien, 'Mechanotransduction and endothelial cell homeostasis: The wisdom of the cell', *American Journal of Physiology: Heart and Circulatory Physiology*, vol. 292, no. 3, March 2007, H1209–24.

20 E. Shokri-Kojori et al., 'An autonomic network: Synchrony between slow rhythms of pulse and brain resting state is associated with personality and emotions', *Cerebral Cortex*, vol. 29, no. 4, 1 April 2019, p. 1702.

21 G.S. Chan et al., 'Contribution of arterial Windkessel in low-frequency cerebral hemodynamics during transient changes in blood pressure', *Journal of Applied Physiology*, vol. 110, no. 4, 2011, pp. 917–25; I. Zamzuri, 'Searching for the origin through central nervous system: A review and thought which related to microgravity, evolution, big bang theory and universes, soul and brainwaves, greater limbic system and seat of the soul', *Malaysian Journal of Medical Science*, vol. 21, no. 4, July 2014, pp. 4–11.

22 Samuels, 'The heart–brain connection'; R. Gordan et al., 'Autonomic and endocrine control of cardiovascular function', *World Journal of Cardiology*, vol. 7, no. 4, 2015, pp. 204–14.

23 S. Romanenko et al., 'The interaction between electromagnetic fields at megahertz, gigahertz and terahertz frequencies with cells, tissues and organisms: risks and potential', *Journal of the Royal Society Interface*, vol. 14, no. 137, December 2017.

24 R. McCraty, 'New frontiers in heart rate variability and social coherence research: Techniques, technologies, and implications for improving group dynamics and outcomes', *Frontiers in Public Health*, vol. 5, 12 October 2017, p. 267; N. Herring & D.J. Paterson, 'Neuromodulators of peripheral cardiac sympatho-vagal balance', *Experimental Physiology*, vol. 94, no. 1, January 2009, pp. 46–53.

25 F. Shaffer et al., 'A healthy heart is not a metronome: An integrative review of the heart's anatomy and heart rate variability', *Frontiers in Psychology*, vol. 30, no. 5, September 2014.

26 B. Vickhoff et al., 'Music structure determines heart rate variability of singers', *Frontiers in Psychology*, vol. 4, no. 334, 9 July 2013.

27 F. Lombardi, 'Chaos theory, heart rate variability, and arrhythmic mortality', *Circulation*, vol. 101, no. 1, 4 January 2000, pp. 8–10; P.C. Ivanov et al., 'Multifractality in human heartbeat dynamics', *Nature*, vol. 399, no. 6735, 3 June 1999, pp. 461–65.

28 D.C. Lin & A. Sharif, 'Common multifractality in the heart rate variability and brain activity of healthy humans', *Chaos*, vol. 20, no. 2, June 2010.

29 Lin & Sharif, 'Common multifractality'; A.H. Kemp et al., 'From psychological moments to mortality: A multidisciplinary synthesis on heart rate variability spanning the continuum of time', *Neuroscience & Biobehavioral Reviews*, vol. 83, December 2017, pp. 547–67.

30 E. Chargaff, *Heraclitean Fire: Sketches from a life before nature*, The Rockefeller University Press, New York, 1978, p. 179.

31 Shaffer et al., 'A healthy heart'.

32 Kemp et al., 'From psychological moments to mortality'.

Heart Tone

1 D. Ladinsky, *A Year with Hafiz: Daily contemplations*, Penguin, New York, 2011.

2 Shaffer et al., 'A healthy heart'; R. McCraty, 'New frontiers in heart rate variability and social coherence research'.

Wisdom from the Heart

1 A.L. Hansen et al., 'Vagal influence on working memory and attention', *International Journal of Psychophysiology*, vol. 48, pp. 263–74; I. Grossmann et al., 'A heart and a mind: Self distancing facilitates the association between heart rate variability, and wise reasoning', *Frontiers in Behavioral Neuroscience*, vol. 10, 8 April 2016, p. 68.

2 Kemp et al., 'From psychological moments to mortality'.

3 A. Lischke, 'Interindividual differences in heart rate variability are associated with interindividual differences in empathy and alexithymia', *Frontiers in Psychology*, vol. 9, 27 February 2018, p. 229.

4 S.U. Maier & T.A. Hare, 'Higher heart-rate variability is associated with ventromedial prefrontal cortex activity and increased resistance to temptation in dietary self-control challenges', *Journal of Neuroscience*, vol. 37, no. 2, 11 January 2017, pp. 446–55.

5 Grossmann et al., 'A heart and a mind'.

6 D. Sinnecker et al., 'Expiration-triggered sinus arrhythmia predicts outcome in survivors of acute myocardial infarction', *Journal of the American College of Cardiology*, vol. 67, no. 19, 17 May 2016, pp. 2213–20.

7 S. Hillebrand et al., 'Heart rate variability and first cardiovascular event in populations without known cardiovascular disease: Meta-analysis and dose-response meta-regression', *Europace*, vol. 15, no. 5, May 2013, pp. 742–49.

8 Thayer et al., 'The relationship of autonomic imbalance, heart rate variability and cardiovascular disease risk factors', *International Journal of Cardiology*, 28 May 2010, vol. 141, no. 2, pp. 122–31.

9 A. Tawakol et al., 'Relation between resting amygdalar activity and cardiovascular events: A longitudinal and cohort study', *The Lancet*, vol. 389, no. 10071, 25 February 2017, pp. 834–45.

10 A.H. Kemp et al., 'Effects of depression, anxiety, comorbidity, and antidepressants on resting- state heart rate and its variability: An ELSA-Brasil cohort baseline study', *American Journal of Psychiatry*, vol. 171, no. 12, 2014, pp. 1328–34.

11 B.H. Friedman, 'An autonomic flexibility-neurovisceral integration model of anxiety and cardiac vagal tone', *Biological Psychology*, vol. 74, 2007, pp. 185–99.

12 U. Kumar et al., 'Neuro-cognitive aspects of "OM" sound/syllable perception: A functional neuroimaging study', *Cognitive Emotions*, vol. 29, no. 3, 2015, pp. 432–41.

13 B. Allen et al., 'Resting high-frequency heart rate variability is related to resting brain perfusion', *Psychophysiology*, vol. 52, no. 2, 2015, pp. 277–87.

Hearts in Sync

1 B.A. Danalache et al., 'Oxytocin-Gly-Lys Arg stimulates cardiomyogenesis by targeting cardiac side population cells', *Journal of Endocrinology*, vol. 220, no. 3, 30 January 2014, pp. 277–89.

2 T. Oyama et al., 'Cardiac side population cells have a potential to migrate and differentiate into cardiomyocytes in vitro and in vivo', *Journal of Cell Biology*, vol. 176, 2007, pp. 329–41; J. Paquin et al., 'Oxytocin induces differentiation of P19 embryonic stem cells to cardiomyocytes', *Proceedings of the National Academy of Sciences of the United States of America*, 2002, pp. 999550–55.

3 Jones et al., 'Ethological observations of social behavior in the operating room'.

4 T. Müller, 'Ethikrat bekennt sich zur bestehenden Hirntod-Praxis' ['Ethics Council commits to existing brain death practice'], *ÄrzteZeitung*, 24 February 2015, www.aerztezeitung.de/politik_gesellschaft/organspende/article/880051/organspende-ethikrat-bekennt-bestehenden-hirntodpraxis.html, accessed 7 December 2018.

5 J.A. Dipietro et al., 'Prenatal development of intrafetal and maternal-fetal synchrony', *Behavioral Neuroscience*, vol. 120, 2006, pp. 687–701; J. Patrick et al., 'Influence of maternal heart rate and gross fetal body movements on the daily pattern of fetal heart rate near term', *American Journal of Obstetrics and Gynecology*, vol. 144, 1982, pp. 533–38; S. Lunshof et al., 'Fetal and maternal diurnal rhythms during the third trimester of normal pregnancy: Outcomes of computerized analysis of continuous twenty-four-hour fetal heart rate recordings', *American Journal of Obstetrics and Gynecology*, vol. 178, 1998, pp. 247–54.

6 P. Van Leeuwen et al., 'Influence of paced maternal breathing on fetal-maternal heart rate coordination', *Proceedings of the National Academy of Sciences of the United States of America*, vol. 106, no. 33, 18 August 2009, pp. 13661–66.

7 P. Ivanov et al., 'Maternal-fetal heartbeat phase synchronization', *Proceedings of the National Academy of Sciences of the United States of America*, vol. 106, no. 33, 18 August 2009, pp. 13641–42.

8 C. Huygens, *Horologium Oscillatorium: Sive de motu pendulorum ad horologia aptato demonstrationes geometricae* [*The Pendulum Clock: Or geometrical demonstrations concerning the motion of pendula as applied to clocks*], Iowa State University Press, 1986 [1673].

9 J.B. Bavelas et al., 'Listener responses as a collaborative process: The role of gaze', *Journal of Communication*, vol. 52, no. 3, 2002, pp. 566–80; A.S. Pikovsky et al., *Synchronization: A universal concept in nonlinear science*, Cambridge University Press, Cambridge (UK), 2001.

10 P. Ostborn et al., 'Phase transitions towards frequency entrainment in large oscillator lattices', *Physical Review*, 2003, E68: 015104.

11 P. Van Leeuwen et al., 'Aerobic exercise during pregnancy and presence of fetal-maternal heart rate synchronization', *PLoS One*, vol. 9, no. 8, 27 August 2014; C. Porcaro et al., 'Fetal auditory responses to external sounds and mother's heart beat: Detection improved by Independent Component Analysis', *Brain Research*, vol. 1101, 2006, pp. 51–58.

12 J. Gutkowska et al., 'The role of oxytocin in cardiovascular regulation', *Brazilian Journal of Medical and Biological Research*, vol. 47, no. 3, 2014, pp. 206–14.

13 Danalache et al., 'Oxytocin-Gly-Lys Arg stimulates cardiomyogenesis'; M. Jankowski et al., 'Oxytocin in cardiac ontogeny', *Proceedings of the National Academy of Sciences of the United States of America*, vol. 101, no. 35, 31 August 2004, pp. 13074–79.

14 J.A. DiPietro, 'Psychological and psychophysiological considerations regarding the maternal-fetal relationship', *Infant Child Development*, vol. 19, no. 1, 2010, pp. 27–38.

15 R. McCraty, *Science of the Heart: Exploring the role of the heart in human performance*, vol. 2, HeartMath Institute, 2015.

16 X. Cong et al., 'Parental oxytocin responses during skin to skin contact with preterm infants', *Early Human Development*, vol. 91, 2015, pp. 401–06.

17 D. Vittner et al., 'Increase in oxytocin from skin-to-skin contact enhances development of parent-infant relationship', *Biological Research for Nursing*, vol. 20, no. 1, January 2018, pp. 54–62.

18 A.S. Kadic & A. Kurjak, 'Cognitive functions of the fetus', *Ultraschall in der Medizin – European Journal of Ultrasound*, vol. 39, no. 2, April 2018, pp. 181–9; E.R. Sowell et al., 'Longitudinal mapping of cortical thickness and brain growth in normal children', *The Journal of Neuroscience*, vol. 24, no. 38, 2004, pp. 8223–31.

19 Merker, 'Consciousness without a cerebral cortex'.

20 R. Brusseau, 'Developmental perspectives: Is the fetus conscious?', *International Anesthesiology Clinic*, vol. 46, no. 3, Summer 2008, pp. 11–23.

21 Kadic & Kurjak, 'Cognitive functions of the fetus'; G.Z. Tau & B.S. Peterson, 'Normal development of brain circuits', *Neuropsychopharmacology*, vol. 35, no. 1, 2010, pp. 147–68.

22 Porcaro et al., 'Fetal auditory responses'; R. Draganova et al., 'Sound frequency change detection in fetuses and newborns: A magnetoencephalographic study', *NeuroImage*, vol. 28, 2005, pp. 354–61; K. Dunn et al., 'The functional foetal brain: A systematic preview of methodological factors in reporting foetal visual and auditory capacity', *Developmental Cognitive Neuroscience*, vol. 13, June 2015, pp. 43–52.

The Heart in the Incubator

1 A. Hemakom et al., 'Quantifying team cooperation through intrinsic multi-scale measures: respiratory and cardiac synchronization in choir singers and surgical teams', *Royal Society Open Science*, vol. 4, no. 12, 6 November 2017, p. 170853.

2 P.M. Lehrer & R. Gevirtz, 'Heart rate variability biofeedback: How and why does it work?', *Frontiers in Psychology*, vol. 5, 2014; Kemp et al., 'From psychological moments to mortality'; Vickhoff et al., 'Music structure determines heart rate variability of singers'.

3 L. Bernardi et al., 'Effect of rosary prayer and yoga mantras on autonomic cardiovascular rhythms: Comparative study', *British Medical Journal*, vol. 323, 2001, pp. 1446–49.

4 J.E. Mathieu et al., 'The influence of shared mental models on team process and performance', *Journal of Applied Psychology*, vol. 85, 2000, pp. 273–83; J.A. Cannon-Bowers & E. Salas, 'Reflections on shared cognition', *Journal of Organisational Behavior*, vol. 22, 2001, pp. 195–202.

5 Vickhoff et al., 'Music structure determines heart rate variability of singers'.
6 Porcaro et al., 'Fetal auditory responses to external sounds and mother's heart beat'.

What the Heart Can Feel

1 Gordan et al., 'Autonomic and endocrine control of cardiovascular function'; Shivkumar et al., 'Clinical neurocardiology'.
2 Herring & Paterson, 'Neuromodulators of peripheral cardiac sympatho-vagal balance'.
3 McCraty, *Science of the Heart*.
4 D. Childre et al., *Heart Intelligence*, San Francisco, Waterfront Press, 2016.
5 J.L. Helm et al., 'Assessing cross-partner associations in physiological responses via coupled oscillator models', *Emotion*, vol. 12, no. 4, August 2012, pp. 748–62.
6 K.C. Light et al., 'More frequent partner hugs and higher oxytocin levels are linked to lower blood pressure and heart rate in premenopausal women', *Biological Psychology*, vol. 69, 2005, pp. 5–21.
7 J. Gutkowska et al., 'Oxytocin releases atrial natriuretic peptide by combining with oxytocin receptors in the heart', *Proceedings of the National Academy of Sciences of the United States of America*, vol. 94, no. 21, 14 October 1997, pp. 11704–09; Gutkowska et al., 'The role of oxytocin in cardiovascular regulation'.
8 I. Schneiderman et al., 'Love alters autonomic reactivity to emotions', *Emotion*, vol. 11, no. 6, December 2011, pp. 1314–21.
9 J. Chatel-Goldman et al., 'Touch increases autonomic coupling between romantic partners', *Frontiers in Behavioral Neuroscience*, vol. 8, 27 March 2014, p. 95; P. Goldstein et al., 'The role of touch in regulating interpartner physiological coupling during empathy for pain', *Scientific Reports*, vol. 7, no. 1, 12 June 2017, p. 3252.
10 S.C. Walker et al., 'C-tactile afferents: Cutaneous mediators of oxytocin release during affiliative tactile interactions?', *Neuropeptides*, vol. 64, August 2017, pp. 27–38; K. Uvnäs-Moberg et al., 'Self-soothing behaviors with particular reference to oxytocin release induced by non-noxious sensory stimulation', *Frontiers in Psychology*, vol. 5, 12 January 2015, p. 1529.

11 Goldstein et al., 'The role of touch in regulating interpartner physiological coupling'.
12 M.H. Huang et al., 'An intrinsic adrenergic system in mammalian heart', *Journal of Clinical Investigation*, vol. 98, no. 6, 15 September 1996, pp. 1298–1303; M.H. Huang et al., 'Neuroendocrine properties of intrinsic cardiac adrenergic cells in fetal heart rate', *American Journal of Physiology Heart Circulation Physiology*, vol. 288, no. 2, February 2005, H497–503.
13 S. Tarlaci, 'The brain in love: Has neuroscience stolen the secret of love?', *NeuroQuantology*, vol. 10, no. 4, 2012, pp. 744–53.
14 T. Maurice & T.P. Su, 'The pharmacology of sigma-1 receptors', *Pharmacology & Therapeutics*, vol. 124, no. 2, November 2009, pp. 195–206; J.M. Beaulieu & R.R. Gainetdinov, 'The physiology, signaling, and pharmacology of dopamine receptors', *Pharmacological Review*, vol. 63, no. 1, March 2011, pp. 182–217; Schneiderman et al., 'Love alters autonomic reactivity to emotions'; M. Leonti & L. Casu, 'Ethnopharmacology of love', *Frontiers in Pharmacology*, vol. 9, 3 July 2018, p. 567; Gordan, 'Autonomic and endocrine control of cardiovascular function'; McCraty, *Science of the Heart.*
15 J.L. Helm et al., 'Assessing cross-partner associations in physiological responses via coupled oscillator models', *Emotion*, vol. 12, no. 4, August 2012, pp. 748–62.

Danse Macabre

1 P. Van Lommel, 'Near death experience, consciousness and the brain: A new concept about the continuity of our consciousness based on recent scientific research on near death experience in survivors in cardiac arrest', *World Futures*, vol. 62, pp. 134–51, 200.
2 P. Van Lommel et al., 'Near-death experience in survivors of cardiac arrest: A prospective study in the Netherlands', *The Lancet*, vol. 358, no. 9298, 15 December 2001, pp. 2039–45; Van Lommel, 'Near death experience, consciousness and the brain'; P. Van Lommel, *Consciousness Beyond Life: The science of the near-death experience*, HarperCollins, New York, 2010.
3 E. Alexander, *Proof of Heaven: A neurosurgeon's journey into the afterlife*, Simon & Schuster, New York, 2012.

4 Dharma University, *Wenn sich der Geist vom Körper löst* [*When the Spirit Leaves the Body*], www.dharma-university-press.org/component/k2/item/33-wenn-sich-der-geist-vom-koerper-loest.html, accessed 16 March 2020.

5 Van Lommel et al., 'Near-death experience in survivors of cardiac arrest'; Van Lommel, 'Near death experience, consciousness and the brain'; Van Lommel, *Consciousness Beyond Life*.

6 R.D. Truog et al., 'The 50-year legacy of the Harvard report on brain death', *JAMA*, vol. 320, no. 4, 24 July 2018, pp. 335–36; A.S. Iltis & M.J. Cherry, 'Death revisited: Rethinking death and the dead donor rule', *Journal of Medicine and Philosophy*, vol. 35, no. 3, June 2010, pp. 223–41.

7 Truog et al., 'The 50-year legacy of the Harvard report on brain death'.

8 J.M. Luce, 'The uncommon case of Jahi McMath', *Chest*, vol. 147, no. 4, April 2015, pp. 1144–51.

9 N. Nayhauss, 'Hirntote Mutter aus Berlin bringt Kind zur Welt und stirbt' ['Braindead mother from Berlin gives birth and dies'], *Berliner Morgenpost*, 27 February 2018, www.morgenpost.de/berlin/article213565437/Wie-eine-Berliner-Mutter-ihr-Kind-bekam-und-starb.html, accessed 16 March 2020.

10 H. Stolp, 'Organspende: Übertragen Organe Bewusstsein?' ['Organ donation: Do organs transmit consciousness?', *Crotana*, 2016.

11 Deutsche Stiftung Organtransplantation [German Organ Transplantation Foundation], www.dso.de, accessed 16 March 2018.

12 Merker, 'Consciousness without a cerebral cortex', D. Shewmon et al., 'The use of anencephalic infants as organ sources: A critique', *JAMA*, vol. 261, 1989, pp. 1773–81.

13 G. Citerio & P.G. Murphy, 'Brain death: The European perspective', *Seminars in Neurology*, vol. 35, no. 2, April 2015, pp. 139–44; E.F. Wijdicks, 'The transatlantic divide over brain death determination and the debate', *Brain*, vol. 135 (Pt 4), April 2012, pp. 1321–31.

14 A. Riepertinger, *Mein Leben mit den Toten. Ein Leichenpräparator erzählt* [*My Life with the Dead: Tales of an embalmer*], Heyne, Munich, 2012.

The Heart in the Eyes

1 H.D. Park et al., 'Spontaneous fluctuations in neural responses to heartbeats predict visual detection', *Nature Neuroscience*, vol. 17, no. 4, 2014, p. 612–18.

2 R. Ryan, *I Thought about It in My Head and I Felt It in My Heart but I Made It wit My Hands*, Rizzoli, New York, 2018.

3 Park et al., 'Spontaneous fluctuations'.

4 H. Fukushima, Y. Terasawa & S. Umeda, 'Association between interoception and empathy: Evidence from heartbeat-evoked brain potential', *International Journal of Psychophysiology*, vol. 79, no. 2, 2011, pp. 259–65; C.D.B. Luft & J. Bhattacharya, 'Aroused with heart: Modulation of heartbeat evoked potential by arousal induction and its oscillatory correlates', *Scientific Reports*, vol. 5, 27 October 2015.

5 M.A. Gray et al., 'A cortical potential reflecting cardiac function', *PNAS*, vol. 104, 2007, pp. 6818–23; J. Terhaar et al., 'Heartbeat evoked potentials mirror altered body perception in depressed patients', *Clinical Neurophysiology*, vol. 123, 2012, pp. 1950–7.

6 J.S. Winston & G. Rees, 'Following your heart', *Nature Neuroscience*, vol. 17, no. 4, April 2014, pp. 482–83.

7 F. Zöllner (ed.), *Michelangelo: Das Gesamtwerk – Skulptur, Malerei, Architektur, Zeichnungen* [*Michelangelo: The complete works – sculpture, painting, architecture, drawings*], Taschen, Cologne, 2007.

8 I. Konvalinka et al., 'Synchronized arousal between performers and related spectators in a fire-walking ritual', *PNAS*, vol. 108, no. 20, 2011, pp. 8514–19.

9 R. Dunbar et al., *Evolutionary Psychology: A beginner's guide*, One World Press, Oxford, 2005.

Loving

1 Tarlaci, 'The Brain in Love'.

2 R. Feldman et al., 'Oxytocin pathway genes: Evolutionary ancient system impacting on human affiliation, sociality, and psychopathology', *Biological Psychiatry*, vol. 79, no. 3, 1 February 2016, pp. 174–84.

3 H. Yusuf, 'Talks about Lust & Desire', 2011, www.youtube.com/watch?v=7HxDWhPpU-Q, accessed 16 March 2020.

4 S. Brody, 'The relative health benefits of different sexual activities', *Journal of Sexual Medicine*, vol. 7, no. 4 (Pt 1), April 2010, pp. 1336–61.

5 R.M. Costa & S. Brody, 'Female sexual function and heart rate variability', *Applied Psychophysiology Biofeedback*, vol. 40, no. 4, December 2015, pp. 377–78; B.R. Komisaruk et al., 'Brain activation during vagino-cervical stimulation and orgasm in women with complete spinal cord injury: fMRI evidence of mediation by the vagus nerves', *Brain Research*, vol. 1024, 2004, pp. 77–88.

6 D.L. Rowland, 'Neurobiology of sexual response in men and women', *CNS Spectrums*, vol. 11, no. 8 (Suppl 9), August 2006, pp. 6–12.

7 Uvnäs-Moberg et al., 'Self-soothing behaviors with particular reference to oxytocin release induced by non-noxious sensory stimulation'.

8 S.C. Walker et al., 'C-tactile afferents'.

9 Brody, 'The relative health benefits of different sexual activities'; Schneiderman et al., 'Love alters autonomic reactivity to emotions'.

10 D. Udelson, 'Biomechanics of male erectile function', *Journal of the Royal Society Interface*, vol. 4, no. 17, 22 December 2007, pp. 1031–47; Costa & Brody, 'Female sexual function and heart rate variability'.

11 Thich Nhat Hanh, *Love Letter to the Earth*, Parallax, Berkeley, CA, 2013.

12 D. Rothenbacher et al., 'Sexual activity patterns before myocardial infarction and risk of subsequent cardiovascular adverse events', *Journal of the American College of Cardiology*, vol. 66, no. 13, 29 September 2015, pp. 1516–17.

13 L. Lange et al., 'Love death: A retrospective and prospective follow up mortality study over 45 years', *Journal of Sexual Medicine*, vol. 14, no. 10, October 2017, pp. 1226–31.

14 S.M.I. Uddin et al., 'Erectile dysfunction as an independent predictor of future cardiovascular events: The Multi-Ethnic Study of Atherosclerosis', *Circulation*, 11 June 2018.

15 E. Maseroli et al., 'Cardiometabolic risk and female sexuality – Part I. Risk factors and potential pathophysiological underpinnings for female vasculogenic sexual dysfunction syndromes', *Sexual Medicine Review*, vol. 6, no. 4, October 2018, pp. 508–24.

16 N. Jovancevic et al., 'Medium chain fatty acids modulate myocardial function via a cardiac odorant receptor', *Basic Research in Cardiology*, vol. 112, no. 2, March 2017, p. 13.

17 J. Weiler, *Duftrezeptor im menschlichen Herzen. Unser Herz kann riechen – aber warum?* [*'Fragrance receptor in the human heart: Our heart can smell, but why?'*], www.laborpraxis.vogel.de/unser-herz-kann-riechen-aber-warum-a-580193, accessed 16 March 2020.

18 Leonti & Casu, *Ethnopharmacology of Love*.

The Lonely Heart

1 Taggart et al., 'Significance of neuro-cardiac control mechanisms'.

2 A.N. Ganesan et al., 'Long-term outcomes of catheter ablation of atrial fibrillation: A systematic review and meta-analysis', *Journal of the American Heart Association*, vol. 2, no. 2, 2013, e004549.

3 Taggart et al., 'Heart-brain interactions in cardiac arrhythmia'; Taggart et al., 'Significance of neuro-cardiac control mechanisms'.

4 Garg et al., 'Depressive symptoms and risk of incident atrial fibrillation'; American Heart Association News, 'Depression could increase risk of serious heart rhythm condition', https://newsarchive.heart.org/depression-increase-risk-afib-serious-heart-rhythm-condition, accessed 16 March 2020.

5 S. Marinkovic et al., 'Nature, life and mind: An essay on the essence', *Folia Morphol* (Warsz), vol. 74, no. 3, 2015, pp. 273–82; Zamzuri, 'Searching for the origin through central nervous system'.

6 D. Paterson & D. Noble, Video Interviews. Cardiology 1: From early heart research to 21st century challenges. Cardiology 2: The interdisciplinary nature of cardiology. Cardiology 3: Reflections on heart and voices from Oxford 2012, www.dpag.ox.ac.uk/research/paterson-group/media/video-interviews, accessed 16 March 2018.

The Big Picture

1 S. Hawking & L. Mlodinow, *The Grand Design*, Bantam Books, New York, 2010.

2 H.P. Dürr & H. Fuß (interviewer), 'Am Anfang war der Quantengeist' ['In the beginning was the quantum spirit'], *P.M. Magazin*, May 2007, http://oelberg.info/download/artikel/Quantengeist.pdf, accessed 16 March 2020.

3 M. Bojowald, *Once Before Time: A whole story of the universe*, Knopf, New York, 2010.
4 S. Hawking, *A Brief History of Time: From the Big Bang to black holes*, Bantam Dell, New York, 1988.
5 M. Kaku, *The Future of the Mind: The scientific quest to understand, enhance and empower the mind*, Doubleday, New York, 2014.
6 M.C. Keith & R. Bolli, '"String theory" of c-kit(pos) cardiac cells: A new paradigm regarding the nature of these cells that may reconcile apparently discrepant results', *Circulation Research*, vol. 116, no. 7, 27 March 2015, pp. 1216–30.
7 H.P. Dürr, 'Vom Greifbaren zum Unbegreiflichen' ['From the graspable to the unfathomable'], in J.S. de Murillo & M. Thurner (eds), *Aufgang. Jahrbuch für Denken, Dichten Musik*, Bd 6, Verlag W. Kohlhammer, 2009.

Heart Consciousness

1 S. Wiens, 'Heartbeat detection and the experience of emotions', *Cognition and Emotion*, vol. 14, no. 3, 2000, pp. 417–27.
2 Y. Terasawa, 'Interoceptive sensitivity predicts sensitivity to the emotions of others', *Cognition and Emotion*, vol. 28, no. 8, 2014, pp. 1435–48.
3 R.M. Piech et al., 'People with higher interoceptive sensitivity are more altruistic, but improving interoception does not increase altruism', *Scientific Reports*, vol. 7, no. 1, 15 November 2017, p. 15652.
4 D. Badoud & M. Tsakiris, 'From the body's viscera to the body's image: Is there a link between interoception and body image concerns?', *Neuroscience Biobehavioural Review*, vol. 77, June 2017, pp. 237–46.
5 S.N. Garfinkel et al., 'Fear from the heart: Sensitivity to fear stimuli depends on individual heartbeats', *Journal of Neuroscience*, vol. 34, no. 19, 7 May 2014, pp. 6573–82.
6 A. Damasio & G.B. Carvalho, 'The nature of feelings: Evolutionary and neurobiological origins', *Nature Reviews Neuroscience*, vol. 14, no. 2, February 2013, pp. 143–52.
7 E.E. Benarroch, 'HCN channels: Function and clinical implications', *Neurology*, vol. 80, no. 3, 2013, pp. 304–10.

8 K. Shivkumar et al., 'Clinical neurocardiology'.

9 McCraty, 'New frontiers in heart rate variability and social coherence research'; I. Timofejeva et al., 'Identification of a group's physiological synchronization with earth's magnetic field', *International Journal of Environmental Research and Public Health*, vol. 14, no. 9, 2017, pp. 2–22; L. Brizhik et al., 'The role of electromagnetic potentials in the evolutionary dynamics of ecosystems', *Ecological Modelling*, vol. 220, no. 16, 2009, pp. 1865–69.

10 A. Alabdulgader et al., 'Long term study of heart rate variability responses to changes in the solar and geomagnetic environment', *Science Reports*, vol. 8, no. 1, 8 February 2018, p. 2663.

11 R.A. Pollack et al., 'Impact of bystander automated external defibrillator use on survival and functional outcomes in shockable observed public cardiac arrests', *Circulation*, 2018.

12 See: www.nobelprize.org/uploads/2018/06/popular-physicsprize2015.pdf, and a more scientific version: www.nobelprize.org/uploads/2018/06/advanced-physicsprize2015.pdf, accessed 23 October 2020.

13 S. Hameroff & R. Penrose, 'Consciousness in the universe: A review of the "Orch OR" theory', *Physics of Life Reviews*, vol. 11, no. 1, 2014, pp. 39–78; S. Hameroff, 'Quantum walks in brain microtubules: A biomolecular basis for quantum cognition?', *Topics in Cognitive Science*, vol. 6, no. 1, 2014, pp. 91–97.

14 H. Dambeck, 'Physik-Nobelpreis 2015. Jäger der Geisterteilchen' ['Nobel Prize in Physics, 2015: Ghost particle hunter'], www.spiegel.de/wissenschaft/natur/physik-nobelpreis-2015-die-neutrino-jaeger-a-1056476.html, accessed 16 March 2020.

Heart Encounter

1 S.R. Steinhubl et al., 'Cardiovascular and nervous system changes during meditation', *Frontiers of Human Neuroscience*, vol. 9, 18 March 2015, p. 145; G.N. Levine et al., 'Meditation and cardiovascular risk reduction: A scientific statement from the American Heart Association', *Journal of the American Heart Association*, vol. 6, no. 10, 28 September 2017, p. ii.

2 WHO Depression, www.who.int/news-room/fact-sheets/detail/depression, accessed 16 March 2020.

3 F. Werfel (trans. Geoffrey Dunlop), *The Forty Days of Musa Dagh*, The Modern Library, New York, 1934.

4 M. Lisson, 'Einst Herzchirurg, jetzt Busfahrer' ['Formerly heart surgeon, bus driver today'], *ÄrzteZeitung*, 1 November 2012, www.aerztezeitung.de/panorama/article/825594/lebenswandel-einst-herzchirurg-jetztbusfahrer.html, accessed 25 November 2021.

Farewell to the Artificial Heart

1 B. Couto et al., 'The man who feels two hearts: The different pathways of interoception', *Social Cognitive and Affective Neuroscience*, vol. 9, no. 9, September 2019, pp. 1253–60.

2 A.J. Barsky et al., 'Palpitations and cardiac awareness after heart transplantation', *Psychosomatic Medicine*, vol. 60, no. 5, September–October 1998, pp. 557–62.

3 F. Tretter et al., 'Memorandum Reflexive Neurowissenschaft 2014', www.exp.unibe.ch/research/papers/Memorandum%20Reflexive%20 Neurowissenschaft.pdf, accessed 25 November 2021.

Homo cor

1 Damasio & Carvalho, 'The nature of feelings'; P. Goldstein et al., 'The role of touch'.

2 Tretter et al., 'Memorandum'.

3 Hawking & Mlodinow, *The Grand Design.*

4 Hawking & Mlodinow, *The Grand Design.*

5 Y.N. Harari, *Home Deus: A brief history of tomorrow*, Harvill Secker, 2016.

6 T. Esch & G.B. Stefano, 'The neurobiology of pleasure, reward processes, addiction and their health implications', *Neuroendocrinology Letters*, vol. 25, no. 4, August 2004, pp. 235–51.

7 Werfel, *The Forty Days of Musa Dagh.*

INDEX

INDEX

INDEX

INDEX

INDEX

About the Author

Dr Reinhard Friedl is an eminent German surgeon who has held thousands of hearts in his hands. He has operated on premature babies and repaired the heart valves of the very old, implanted artificial heart turbines and stitched up stabbing wounds to the heart.

Shirley Michaela Seul is a professional author.

About the Translator

Dr Gert Reifarth teaches German at Scotch College, Melbourne. Born in the former East Germany, he has also worked in academia and theatre.